SOUND
AND FURY

SOUND AND FURY

The Science and Politics of Global Warming

Patrick J. Michaels

CATO INSTITUTE
Washington, D.C.

Library of Congress Cataloging-in-Publication Data

Michaels, Patrick J.
  Sound and fury : the science and politics of global warming / Patrick J.
  Michaels.
      p.   cm.
  Includes bibliographical references.
  ISBN 0-932790-90-9 : $21.95.—ISBN 0-932790-89-5 : $11.95
  1. Global warming. I. Title.
QC981.8.G56M54   1992
363.73'87—dc20   92-36264
                                                                       CIP

Cover Design by Colin Moore.

Printed in the United States of America.

CATO INSTITUTE
224 Second Street, S.E.
·   Washington, D.C. 20003

Honesty is the first chapter in the book of wisdom.

—Thomas Jefferson

The whole aim of practical politics is to keep the populace alarmed—and hence clamorous to be led to safety—by menacing it with an endless series of hobgoblins, all of them imaginary.

—H. L. Mencken

Contents

PREFACE: A PERSONAL NOTE   ix

MAY 12, 1992   xiii

1. THE VIEW FROM MONTICELLO   1

2. ENHANCING THE GREENHOUSE   9

3. THE INTERGOVERNMENTAL PANEL ON CLIMATE CHANGE   25

4. COMPUTER MODELS: CALCULATING THE APOCALYPSE   31

5. TAKING THE EARTH'S TEMPERATURE   43

6. GHOSTS IN THE APOCALYPSE MACHINE   81

7. CLIMATE CHANGE: THE NEW VISION   89

8. THE NEW VISION: VERIFICATION BY THE DATA   111

9. MEASURING CLIMATE'S IMPACT   131

10. POLITICAL SCIENCE   137

11. COMPETING APOCALYPSES: GLOBAL WARMING, OZONE DEPLETION, AND ACID RAIN   163

12. NEWER CLIMATE MODELS   169

13. CONSENSUS?   181

14. THE LONG-RANGE FORECAST   187

REFERENCES   191

Preface: A Personal Note

This volume grew out of a series of lectures that are part of a class I offer on science, politics, and the environment. Year after year, the material presented here engenders the same question after approximately 15 lectures: "If the data-driven argument against environmental apocalypse from global warming seems so compelling and convincing, why have so few people heard it?"

I don't know. Every piece of technical information provided here is in the refereed scientific literature, is a subsidiary calculation based on that literature, or has been presented at scientific meetings for which presentations were prescreened by a program committee.

On June 6, 1991, an analogous question arose in a Cato Institute debate with Christopher Flavin, research director for the Worldwatch Institute. The event was formally structured: each debater was allowed 20 minutes to present his case and 10 minutes for rebuttal, followed by questions from the floor. Much of the material used in the expository portion is given in this book. At the end of the debate, the last question from the floor was directed to Chris: "Please be specific; was there anything that your opponent presented that was factually incorrect or wrong?" After thinking for a few seconds, Mr. Flavin answered with an unqualified no.

We scientists really are not that different from everyone else. We spend our lives trying to form an internally consistent picture of the world around us. In my chosen occupation—professing ecological climatology as part of a university faculty—I was compelled to attempt that process with respect to the enhanced greenhouse effect, popularly known as "global warming." This book describes that attempt.

A critical event prompted my investigations. In October 1983 two of the (then) three major television networks carried lead stories about a dramatic press conference at which Environmental Protection Agency spokesman John Hoffman had announced that unprecedented climatic changes would be on us in a very few years, that sea levels would rise enough to wipe out the downtowns of several major cities, and that hundreds of scientists had reviewed and approved the agency's findings.

What I witnessed was the first grand attempt at environmental science (and therefore policy) by press release, a process that has since become numbingly repetitive. I knew then that there had been precious little warming of the planet in the past 100 years, especially in the Northern Hemisphere, and that the EPA's position could not be based on data.

I also sensed that the issue was sufficiently emotional that people who did not go along might be subject to rough treatment. In what is an unfortunate commentary on modern times, because of fear of losing my position, I did not write anything on the subject in a public venue until 1986, two weeks after I was promoted to associate professor. At that time I naively believed that my half-page piece in the "Outlook" section of the June 15, 1986, *Washington Post* would be viewed as an entertaining and enlightening document that would alert the public to the complicated nature of the problem of global warming, along with the apparent inconsistency between the observed data and apocalyptic projections. Just to make sure, I read the article over the telephone to James Hansen, director of the Goddard Institute for Space Studies of the National Aeronautics and Space Administration. He said it was fine.

I was genuinely taken aback when, within one day, my simple and obvious piece prompted a great deal of personalized invective at an EPA meeting. Sen. Albert Gore, Jr. (D-Tenn.), to whom I gave generally favorable reviews, wrote a nasty little attack in the July 7 *Post*, and I began to realize that people were, shall we say, touchy about the issue. I was further dismayed and personally offended when Gore told *Time* magazine, for its 1989 "Planet of the Year" issue, that my (apparently few) friends and I were responsible for the destruction of the planet. According to Gore, "That we face an ecological crisis without any precedent in historic times is no longer a matter of any dispute worthy of recognition," and "those who, for the purpose of maintaining balance in the debate, take the contrarian view that there is significant uncertainty about whether it's real are hurting our ability to respond."

And finally, in a March 18, 1989, *New York Times* piece, Senator Gore analogized those who argue for caution (a data-based argument that forms the core of this book) to people who permitted (by ignoring the signs) the murder of millions by Nazi Germany. That analogy was personally repugnant because I had many relatives who experienced that trauma.

There has been a constructive side to all the disturbance: a strong argument has emerged that the climate apocalypse is hardly at hand. I do not think that view would have acquired such strength without the attacks that pummeled the people who initially presented it. In fact, the anti-apocalyptic argument is now so strong that it is becoming mainstream. See, for example, Gregg Easterbrook's article, "A House of Cards," in the June 1, 1992, issue of *Newsweek* or the three-page *Washington Post* summary of the science of climate change that appeared on May 31, 1992, before the opening of the "Earth Summit."

For what it is worth, I am internally dogged by the following question: "What if I am wrong?" My only reply is that this book forms the most internally consistent summation that I can achieve, given the observed data and some projections. If my conclusions turn out to be totally wrong, it will mean that the observed data are the most accomplished compulsive liars in the history of science. I did not make them up, and I doubt whether I am creative enough to build a set of numbers that could paint such a picture.

At any rate, if all that is contained herein is dead wrong, that error should become pretty apparent in the next 10 years. Scientists from all portions of the greenhouse spectrum—from theoretical modelers to empiricists tracking the data—have acknowledged that the 1990s will tell the story. It is somewhat reassuring to note that modeler Michael E. Schlesinger of the University of Illinois indicated as recently as August 1992 that the 10-year window will be relatively inconsequential with respect to any overall change, and empiricist Robert Balling of Arizona State University, in a book published the same year, made the same argument from the perspective of observed data.

I apologize to readers who may take offense at my somewhat catty writing style, but I am beyond the age of being able to learn new rhetorical tricks. I first developed my style in the competitive academic camaraderie of the University of Chicago, and I have honed it to a low art in 12 years of publishing the popular *Virginia Climate Advisory*, a quarterly billed as the nation's only "magazine of climate and humor." Readers who can endure, or perhaps even enjoy, this book may want ask for a (free) subscription.

The original title of this manuscript was *The Satanic Gases: Global Warming and Political Science,* an obvious play on Salman Rushdie's *The Satanic Verses.* It was suggested by Bruce Hayden, a fellow climatologist at the University of Virginia, because the enhanced greenhouse gases are popularly thought to make it hotter than Hades, and also because the book's contents are likely to be offensive to practitioners of the most popular new religion to come along since Marxism: apocalyptic environmentalism.

May 12, 1992

Embargoed for release: 2:00 pm, May 12, 1992
Wilbur A. Steger, PhD
CONSAD Research Corporation
Pittsburgh, Pennsylvania
Title: JOBS AT RISK: SHORT-TERM AND TRANSITIONAL EMPLOYMENT IMPACTS OF GLOBAL CLIMATE POLICY OPTIONS.

WASHINGTON, D.C., May 12, 1992—At least six hundred thousand workers in America's most basic industries will lose their jobs if a carbon tax or other policies are enacted to reduce carbon dioxide emissions, according to a new study released today by CONSAD Research Corporation. A carbon tax could result from pending federal legislation.

Designed to stabilize carbon dioxide emissions at 1990 levels by the year 2000, a carbon tax would produce annual losses in Gross National Product of 1.7 percent through the year 2020 and put the U.S. economy into a deep-freeze. . . .

According to Dr. Steger's projections, even in the short run (three to five years), American job losses could surpass 360,000. That job loss would be accompanied with adverse economic conditions, including high inflation, resembling the energy shocks of the 1970's. . . .

Thousands of businesses are likely to confront the choice of closing plants or going into bankruptcy if proposals for a so-called carbon tax are adopted. . . .

Steger also predicts that if a carbon tax is adopted, the jobs of nearly five million American workers will be put "at risk." In addition to an increase in American unemployment, that means that substantial numbers of other workers will experience tangible adverse changes in their terms of employment: reductions in wages and hours worked, increased frequency and duration of layoffs, and diminished prospects for job growth. . . .

A number of recent economic studies have unanimously concluded that imposition of a carbon tax for stabilizing or reducing carbon dioxide emissions would reduce the U.S. standard of living.*

*The mean warming projected by computer forecasts that form the basis for this policy is 4.2°C (7.6°F) for a doubling of atmospheric carbon dioxide. If the policy detailed above is enacted, the mean surface temperature of the planet will be a quarter of a degree cooler in 2030 than it would be if the policy were not adopted. No one would notice the difference.

1. The View from Monticello

> A change in our climate, however, is taking place very sensibly. Both heats and colds are become much more moderate within the memory of even the middle-aged. Snows are less frequent and less deep. They do not often lie, below the mountains, more than one, two, or three days, and very rarely a week. They are remembered to have been formerly frequent, deep, and of long continuance. The elderly inform me, the earth used to be covered with snow about three months in every year. . . .
>
> The eastern and southeastern breezes come on generally in the afternoon. They formerly did not penetrate far above Williamsburg. They are now more frequent at Richmond and every now and then reach the mountains. . . . As the land becomes more cleared, it is probable they will extend still further westward.
>
> —Thomas Jefferson
> *Notes on the State of Virginia*

Jefferson was completely wrong in his assertion that deforestation would enhance the sea breeze, but his was the first statement by an American man of letters that human economic activity could cause large-scale changes in climate.

He was so concerned that he wrote further that an individual should be appointed in each state to keep track of the temperature, and that those individuals should meet every 20 or 30 years or so to see if, in fact, the climate was changing. Things have changed some since Jefferson's time; we now meet every few days. My calendar for the past 12 months included stops in Annapolis, Myrtle Beach, Reno, London, Williamsburg, Detroit, Denver, Baltimore, Orlando, Syracuse, San Francisco, and Washington, D.C., to talk about climatic change, especially global warming.

Along the way, Jefferson's *Notes* made fascinating reading, if only for the observational power of the author of the Declaration of Independence. He describes what grows where, where each river

1

goes, how far upstream each is navigable, what trees grow along-side, and what animals live where. Modern deconstructionists would argue that the *Notes* are nothing but a guide to economic exploitation of the American wilderness. But they were much more, for Mr. Jefferson was also a keen observer of the climate—having taken readings twice a day at Monticello for seven years—and he knew that all the plants and animals grew and thrived within certain climatic limits. Jefferson's *Notes*, in their time, were the best available text on North American ecology.

Surely, as he took his temperatures, measured his rainfall, and received complaints that the British were stealing barometers, he must have thought that a change in climate would change the distribution of plants and animals. It is, and was, as obvious as the biotic panoply so visible from Monticello Mountain. Below that mountain—along what is now Interstate 64—pine and oak dominate. Temperatures on summer afternoons regularly reach the mid-90s, and even though the land receives almost four feet of rain a year, those plants struggle with moisture stress almost every warm day. If we climb toward Afton Mountain and the Blue Ridge, hard-woods increase in frequency, resulting in a brilliant display that peaks in late October. The reason the area looks more like New England than Richmond (Jefferson noted that the same species that grow in New Hampshire seem perfectly at home in the Virginia mountains) is that growing season temperatures are about the same in the two regions. On top of the Blue Ridge high winds stunt the trees, and they are subject to damaging ice storms with a frequency that is an order of magnitude greater than the frequency of such storms in the lowlands. The freeze-free period is some two months less than it is in the hardwood zone, only 1,500 feet beneath the crest.

Although he had neither the time nor the instrumentation to measure all those phenomena, Jefferson knew how sensitive trees are, for he observed that those in middle elevations changed color much later than trees at the top of the mountain or down on the piedmont, where cold air settles in the low spots and along the creeks and rivers. Jefferson—whose writing far outshines that of almost all of today's politically active scientists (Jefferson was one, too)—never said that human beings should adjust their economic activity because it apparently has a propensity to change the climate. What he did say was "study it."

There is no longer any significant disagreement in the scientific community that the greenhouse effect is real and already occurring; the resultant increase in temperatures will measure several degrees.

—Sen. Albert Gore, Jr., July 1986

The unequivocal detection of the enhanced greenhouse effect from observations is not likely for a decade or more.

—Policymakers Summary, UN
Intergovernmental Panel on Climate
Change, July 1990

According to most climatology texts, as well as to the landmark UN report of the Intergovernmental Panel on Climate Change (IPCC), the current greenhouse effect from water vapor, carbon dioxide, and other gases is 33°C, or 59°F. Without those gases, the mean temperature of the planet would be inhospitable to life as we know it. That truism, which begins most arguments about global warming and climatic change, is profoundly misleading because it assumes that the earth would look the same without water in its atmosphere (i.e., it would somehow still have oceans and clouds). If you remove the clouds and oceans, which must be done to take water vapor out of the atmosphere, the temperature is not that much different than it is today. Thus, most arguments about global warming begin with a distortion.

Because the surface of the sun is so hot, much of its radiation is in the visible and ultraviolet portions of the electromagnetic spectrum—the short, energetic wavelengths that cause sunburn and skin cancer. Because of its distance from the sun, the earth's surface is much colder, and the laws of physics require that it radiate primarily longer infrared wavelengths. Those less energetic wavelengths—no one gets earthburn—warm mostly the lower atmosphere. Several natural molecules, notably water and carbon dioxide, absorb the infrared radiation and redirect some of it back down toward the earth's surface, which has the effect of warming the lower levels of the atmosphere even further. That is the greenhouse effect, and if it alone were responsible for our mean temperature, the planet would average around 77°C (170.6°F) and be far too hot to sustain what we think of as life.

3

While at Monticello, Jefferson took the temperature twice a day for seven years (1810–16). His recorded mean temperature was approximately 1.0°F (0.6°C) colder than it is today, or about the same temperature as was recorded when official records began about a hundred years ago.

In fact, other atmospheric processes governed by the laws of thermodynamics "short circuit" the greenhouse effect by some 62°C (111.6°F), and those processes are responsible for life as we know it—perhaps more responsible than the vaunted greenhouse effect. Natural cooling of greenhouse warming will occur whether or not man puts more greenhouse gases into the atmosphere.

The existence of the natural greenhouse effect has been known for more than 100 years. In fact, the capacity of carbon dioxide to absorb infrared radiation has been known ever since British physicist John Tyndall's pioneering experiments in the 1870s, but the climatic implications were not much appreciated until they were elucidated by Svante Arrhenius in an 1896 paper in *Philosophical Transactions* in which he calculated that a doubling of the atmospheric CO_2 concentration would result in a net global surface warming of approximately 5°C.

That reference pops up again and again in both the popular and the scientific literature, and it is cited as evidence of the robustness of the upcoming climate apocalypse glibly proclaimed by environmentalists. Such selective quoting neglects Arrhenius's accompanying calculation that for a 50 percent increase in atmospheric CO_2 (which we have effectively reached already—more on that later), the mean temperature should have risen over 3°C . The following passage, which appears slightly later in Arrhenius's paper, is rarely, if ever, quoted.

> The influence is in general greater in the winter than in the summer, except in the case of the parts that lie between the maximum and the pole. The influence . . . is in general somewhat greater for land than ocean. On account of the nebulosity of the southern hemisphere, the effect will be less there than in the northern hemisphere. An increase in [CO_2] will of course diminish the difference in temperature between day and night. A very important . . . secondary effect will probably remove the maximum effect from lower parallels to the neighbourhood of the poles.

As we shall see, that type of warming—concentrated mainly in the high latitudes and occurring mainly in winter—is hardly a basis for an apocalyptic vision and, in fact, is more likely the opposite.

On June 23, 1988, NASA scientist James E. Hansen, head of the Goddard Institute for Space Studies, testified to both the House of Representatives and the Senate that "global warming is now sufficiently large that we can ascribe with a high degree of confidence a cause and effect relationship to the greenhouse effect." Judging from the succeeding issues of *Science* magazine, many environmentalists but few climatologists agreed with that view. Further, Hansen did not say that the heat and drought in the Midwest were caused by human alterations of the greenhouse effect, but he might as well have. Two days later CNN ran a yes/no telephone poll, and 70 percent of the respondents answered yes to a statement that the drought was caused by the greenhouse effect.

CNN had uncovered the "Popular Vision" of global warming caused by the enhanced greenhouse effect—a rapid planetary warming of approximately 4°C (7°F) with a major sea level rise (up to 25 feet, according to some pronouncements made in the early 1980s) caused mainly by the melting of major areas of land ice, especially in Greenland and Antarctica. Also in the Popular Vision is withering corn, as daytime temperatures regularly exceed 100°F (38°C) in the nation's heartland, and climate changes so rapid that forests will be unable to move northward at the pace of the temperature increase, which will result in massive deforestation and desertification. All of those changes will take place while population increases rapidly, so war over the earth's rapidly depleting resources seems to be highly probable. The Popular Vision is nothing short of apocalyptic. If 70 percent of the population subscribes to it, no elected official (even in Washington, D.C., the incumbency capital of Western democracies) can ignore it.

Solutions are expensive: the Electric Power Research Institute estimates that the cost of a serious attempt to reduce enhanced greenhouse emissions is in excess of $5 trillion and would entail annual GNP reductions in the United States alone of more than 2 percent. It is not much of a surprise that Lester Brown of Worldwatch Institute has written that "getting on the path [to prevent climatic change] depends on a wholesale reordering, a fundamental restructuring of the world economy." Although University of Maryland economist Julian Simon says Brown's track record at environmental prognostication is none too good, he is probably right about this one: any attempt to stabilize the greenhouse effect would require a 60 to 80 percent reduction in the emissions of CO_2, the gas that, at least until now, has been the by-product of civilization.

In the absence of a breakthrough in some other form of dense energy (e.g., fusion power), and with the massive resistance to fission (conventional nuclear) power that occupies the same political niche as global warming, any cleanup effort can surely destroy the economic engine of Western industrialized democracies. Further, such a reduction of emissions is simply unattainable without levels of command, control, and tax coercion that will be abhorrent to most Americans.

All of those concerns are unwarranted for one reason: the Popular Vision is wrong. The most internally consistent case that can be generated from billions and billions of bits of climatic data simply does not support that vision.

In this book I will show why the Popular Vision is erroneous and demonstrate that the climatic history of the planet is inconsistent with forecasts of doom and gloom. I will examine the many signs that indicate that the opposite is occurring: We are creating a world in which the winters warm and the summers do not, a world in which the nights warm and the days do not. We are creating a world in which the growing season lengthens and the great ice fields of Greenland and Antarctica change little (they may even be enlarging). The CO_2 we are emitting to the atmosphere has an additional effect: when plants are supplied adequate nutrients, they grow better.

If the data are correct, the likelihood that we are creating a better world far outweighs the probability of climate apocalypse.

So why did not the observant Jefferson, who knew so well the climatic habitat of the otter, the grouse, and his beloved dogwood, speak as glibly as we do today about "radically restructuring" society for the sake of the environment? Perhaps because he believed the government that governs least governs best.

It is very likely that if Jefferson were president today, he would be calling for an intensive investigation of the problem of climatic change and how it might affect ecological change. He would state that our policy should be commensurate with our science.

2. Enhancing the Greenhouse

The quotation from Senator Gore in the previous chapter—"there is no longer any significant disagreement in the scientific community that the greenhouse effect is real"—is about as profound as a revelation that all scientists now agree that the earth is round. The implication and the connection that most people make are that all scientists agree that global temperatures are rising disastrously as a result of the enhancement of the greenhouse effect. That conclusion is simply wrong.

The United Nations' 1990 report of the Intergovernmental Panel on Climate Change (IPCC), often cited as the consensus of climatologists, leads off with a similarly profound revelation: "There is a natural greenhouse effect that already [why is this word necessary?] keeps the earth warmer than it would otherwise be." S. Fred Singer, who is on leave from the University of Virginia and working with the Science and Environmental Policy Project, found that 40 percent of the scientists who contributed to the UN report believed that the statement might convey a misleading message to the public that disastrous warming is "already" at hand. Nearly 100 percent of an equally prestigious group of scientists who did not work on the report felt the same way.

Greenhouse gases warm the lowest layers of the atmosphere—where everything lives—by redirecting radiation that would normally escape directly into space. The most common of those is water vapor, and its concentration does not change very much. Next most important, with about one-seventh of the warming potential of water vapor, is CO_2, but its concentration has hardly been constant throughout the earth's history.

If all the carbon dioxide were removed from the earth's atmosphere, the drop in surface temperature would be 1.5°C (2.7°F), or only 5 percent of the total warming caused by all of the greenhouse gases. That is because the warming caused by those gases does not add up in a simple linear fashion.

In fact, the temperature response to those gases is far from simple and varies with their concentration. Temperatures are quite sensitive to small increments of CO_2 as long as the initial concentration is low. As concentrations increase, the response becomes muted, and eventually the temperature does not change. For example, the amount of warming that should occur for the 33 percent increase in CO_2 that we have already induced in the atmosphere should be more than would occur for the next increment of the same magnitude.

Hundreds of millions of years ago the sun was considerably dimmer, emitting only 80 percent of its current radiation. If everything else were equal, that reduced radiation would have driven the temperature of the earth's surface to freezing, most likely icing the planet forever. But as a result of the complex chemistry of the earth and its atmosphere, greenhouse gas—probably water vapor—concentrations were much higher at that time, and the earth's temperature was not appreciably different than it is today.

In fact, throughout most of the past billion years, the CO_2 concentration of the atmosphere has been greater than it is today. The same is also true for most of the past 100 million years, which is the period during which most of our food and fiber crops evolved. Only since the beginning of the ice ages, some 5 million years ago, have temperatures and atmospheric CO_2 fallen to current levels. When it was really cold, at the height of the ice ages (the last advance terminated only 18,000 years ago), the concentration of CO_2 fell to values that were within a hundred million parts per million (ppm) of being unable to support life. Thus, from the perspective of both geological and evolutionary history, the atmosphere is currently impoverished in CO_2. An additional historical peculiarity is that gas bubbles trapped in Antarctic ice tell us that the temperature dropped *before* the CO_2 concentration changed, not after.

Plants take in CO_2 and fix it in the form of carbohydrates in their roots, stems, and leaves. CO_2 in current concentrations is what is known as a "limiting nutrient": there is currently so little of it in the atmosphere that plants cannot get enough. Increasing the concentration increases the growth of almost all plant species, and both laboratory and field experiments have demonstrated that plants flourish as CO_2 concentration goes up. Further, there is no doubt that human industrial activity has increased the CO_2 concentration in the atmosphere and that most (85 to 95 percent, depending on the estimate) of that increase is a result of burning fossil fuels—the same carbon that was deposited in the earth's crust when CO_2 was in excess. The increase is global, even though the lion's share is emitted in the industrial Northern Hemisphere. CO_2 lasts so long in the air (at least 50 years) that it spreads everywhere, including to Antarctica where bubbles trapped in the ice tell us that the concentration was approximately 270 ppm before the Industrial Revolution. Although those measurements are quite tricky—requiring careful analyses of minute quantities of gases that have been trapped for hundreds of years in ice—270 ppm is approximately the background concentration we arrive at if the observed values (taken since 1958 at the Mauna Loa observatory in Hawaii) are extrapolated backwards.

The Mauna Loa history (Figure 2.1) is the only continuous record of directly measured CO_2 concentrations that goes back three

Figure 2.1
MAUNA LOA RECORD OF CO_2 CONCENTRATIONS, 1958–89

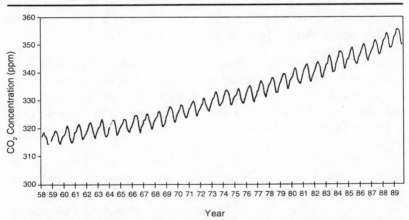

NOTE: The Mauna Loa record of carbon dioxide shows both the overall increase caused primarily by the burning of fossil fuel and the interannual swing in which the planet breaths in CO_2, fixes it in the form of deciduous plant matter, and then "exhales" it back into the atmosphere when the leaves decompose.

11

decades. It has been analyzed very carefully and samples have been tested to see how much of the increase is the result of burning fossil fuel. Tests indicate that approximately 90 percent of the increase is due to that source and that the fraction of CO_2 in the atmosphere as a result of industrial activity is increasing.

Figure 2.1 even shows an annual cycle as trees capture CO_2 and store it in the form of carbohydrates, many of which are discarded each year as leaves fall in the great temperate forests of the Northern Hemisphere. As leaves are shed and ultimately decompose, an increment of CO_2 is released back into the atmosphere, only to be recaptured during the next growing season. That annual cycle represents the breathing of the photosynthesizing biosphere, which requires CO_2 for life.

The breathing of the earth has begun to deepen as a result of increasing concentrations of CO_2; one interpretation is that the planet is becoming greener. Statistically speaking, there is only a 3 percent chance that this is not taking place. The finding, which indicates that plants are taking more CO_2 than they did, should not be surprising, because the atmosphere is merely returning to CO_2 levels that were characteristic during the evolutionary history of almost all terrestrial plants. In a later section of this book I will detail some remarkable experiments that the U.S. Department of Agriculture is currently conducting on that phenomenon.

The current concentration of atmospheric CO_2 is approximately 357 ppm, which represents a rise of one-third from the preindustrial background and about 25 percent over the past 100 years. But the effective increase has been considerably greater because of other infrared-absorbing gases that are also being emitted (see Figure 2.2).

Those other emissions include methane (CH_4), another greenhouse gas whose prime sources include bovine flatulence (cow belching, from either end) and rice paddy agriculture, as well as termites (gassy digestive tracts) and deforestation. The concentration of CH_4 has increased to the point that its effect is equivalent to an additional 30 ppm of CO_2. Certain oxides of nitrogen—thought to be a result of high-intensity agriculture—contribute a few more ppm, and chlorofluorocarbons (CFCs)—which did not even exist before 1950—are good for about 20 ppm.

Figure 2.2
RELATIVE GREENHOUSE WARMING POTENTIAL OF EMISSIONS

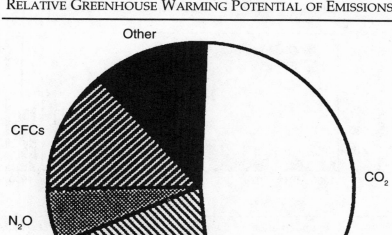

NOTE: According to the Environmental Protection Agency, these are the relative greenhouse warming potentials of various emissions.

In fact, as a result of all the infrared-absorbing emissions, the effective CO_2 concentration is not 357 ppm but 432 ppm (Figure 2.3). That increase is attributable to the contribution of additional methane (30 ppm if it were CO_2), nitrogen oxides (approximately 10 ppm), CFCs (20 ppm),[1] and other emissions (approximately 15 ppm), which, added to 357 ppm, results in a total of 432 ppm, which is 60 percent greater than the concentration was before the emissions that accompanied the industrialization of the planet. And

1. Recent research indicates that the "net" effect of CFCs may be a slight cooling because of their putative destruction of stratospheric ozone. However, cooling would obviously occur only where and when there were large ozone reductions: over Antarctica during late winter and early spring; over other parts of the earth, the warming effect of CFCs would predominate.

Figure 2.3
EFFECTIVE CO_2 CONCENTRATIONS, 1900–90

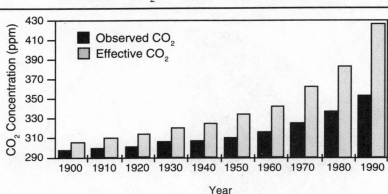

NOTE: The combined effect of the non-CO_2 greenhouse enhancers results in an effective concentration that is approximately 50 percent over the preindustrial background of 270 ppm.

therein lies one of the most intriguing mysteries in science: if greenhouse enhancement invariably leads to an increase in surface temperature, where's the warming?

1988: Birth of the Greenhouse Panic

Journalists loved it. Environmentalists were ecstatic.

—Stephen H. Schneider
National Center for Atmospheric Research

January 1988: Arctic outbreaks chilled the nation and California was dry. Neither fact was unusual. Over the eastern two-thirds of North America, the biggest obstacle to southward-moving cold air is nothing more than a barbed wire fence, so it is the aberrant winter that does not see a big freeze. California has always known feast or famine of winter rain. That condition will prevail as long as there is a jet stream.

Around the first day of spring 1988, precipitation in the corn and soybean belt—a band stretching from eastern Nebraska to southern Ohio—began to taper off, and the succeeding three-month period was as dry as any observed in the 20th century at some locations.

That event was hardly global, or even national, as rain was abundant—too much so—over the winter wheat region to the west and over much of the easternmost portions of the United States. But spring disking and planting raised ominous clouds of dust from Omaha to Cincinnati, and much of what went into the ground did not germinate, which created a temporary local climatic change that dwarfed any global effect.

Before the westward migration (guided, in part, by Jefferson's *Notes*), the corn and soybean region of the United States was a prairie of tall grass—a lush, diverse ecosystem dominated by grasses and spectacular wildflowers. Few people have ever even heard of that prairie, much less seen it, because it no longer exists. The prairie was not just chipped at like the tropical rain forest (one of the few biotic communities that is more diverse), nor was it modified and managed like the great forests of southeastern North America. Instead, it was eradicated, except for a few museum pieces restored at places like the University of Wisconsin's arboretum. A springtime comparison of the arboretum and neighboring cornfields illustrates the dramatic impact of the plow.

Prairie covered the earth with a thick mat, green in the summer, brown in the fall and winter, but it always shielded the black soil from direct sunlight and buffered the local temperature regime. Even in dry years, the deep roots of prairie plants transpired water, which helped to hold down the afternoon heat. In most years, agriculture has the same effect. Corn is planted in late April or May, and before things warm up too much, the leaves mesh, forming a canopy that intercepts most of the incoming solar radiation. The growing plants effectively transpire water; their roots reach down into the soil far beyond where the warmth of the sun could penetrate to cause evaporation. Just as the prairie plants did, today's plants raise the local humidity so much that most of the sun's energy goes into evaporation of water rather than direct heating of the atmosphere.

The manmade prairie failed in 1988 in a way that could not have happened before the plow. Because moisture was so scarce at planting time, few corn or soybean plants germinated, and many of the survivors withered before the canopy closed. Thus, the sun had free access to the flat black surface of the earth, which heated

15

like a parade ground, far beyond a temperature of a "naturally" vegetated surface. The effect was evident in satellite measurements as early as June, and it continued until the fall. Thus, the plow, not CO_2 from automobiles and industry, blew temperatures sky high in the Midwest, evaporating what little water remained and further reducing the supply of moisture available for precipitation. The first half of 1988 was also very warm over many of the planet's other land surfaces, which, while they represent only 32 percent of the total surface of the earth, contain most of the weather stations that have tracked the climate for the past 100 years.

It was big news—the lead story—when James Hansen testified on June 23, 1988, about the "high degree of confidence" that the current climate was related to the enhanced greenhouse effect.

Hansen's testimony received considerable criticism from his scientific peers, as noted in *Science* articles by Richard Kerr that appeared almost monthly. But those comments—too long for "sound bites" and not as flashy as the ubiquitous footage of dead chickens and tractors lost in the dust—were never heard. The real travesty was that no balancing testimony was given.

Hansen, as writer of congressional testimony, was both author and editor—a very dangerous position that abrogates the normal review process. If the issue in question had not been ambiguous—as is the entire global warming process—the story would not have caused a problem. But, as is the case with so many ambiguous issues, balance should have been provided by inviting witnesses to provide counterpoint to strident testimony. That balance did not occur on June 23, 1988, and its lack has remained a hallmark of the entire debate.

In spite of the lack of scientific peer review, the press treats congressional testimony as if it were ex cathedra. Journalists should be more skeptical about that type of information than about any article in the refereed scientific literature, but Hansen's pronouncement received the kind of coverage that is usually reserved for sex scandals.

Was Hansen's testimony really less than balanced? Was it an attempt to galvanize Congress into action? Read his commentary in the *Washington Post* of February 11, 1989: "The evidence for an increasing greenhouse effect is now sufficiently strong that it would have been irresponsible if I had not attempted to alert political leaders."

The main graphic accompanying Hansen's congressional testimony and reproduced in Figure 2.4 (after redigitizing the data for print clarity) was NASA's global temperature history. As we shall see later, that history is hardly a reliable record; it includes data that are obviously in error. Further, Hansen cited an "observed warming during the past century of from 0.6°C to 0.7°C [1.0°F to 1.3°F]," which is 30 percent greater than that in any objective trend analysis of other global temperature records. It was later revealed that he arrived at his figure by subtracting the mean of the first 10 years of the record from the mean of the last 10 years and ignoring the intervening 80 years of data. Throwing out 80 percent of the data to make a striking pronouncement hardly seems to be normal scientific procedure.

Figure 2.4
NASA's GLOBAL TEMPERATURE RECORDS, 1880–1988

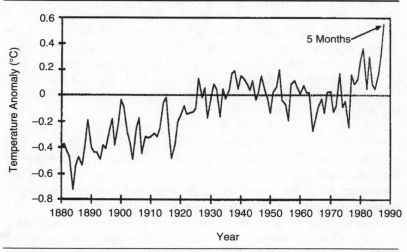

NOTE: The temperature history that NASA scientist James Hansen presented to Congress on June 23, 1988, mixed scientific "apples and oranges" by comparing a five-month sample of data to annual averages. To appreciate the biasing effect of the last point, you should cover it with your finger.

Nonetheless, the record shows something that has been known in the climatological community for some time: the earth's land-masses were colder in the 19th century, and a fairly rapid rise in global mean temperature occurred before World War II—that is, before most of the anthropogenic greenhouse gases were even put into the atmosphere. Therefore, the timing of much of the observed warming is not consistent with an enhanced greenhouse effect. Since then, as greenhouse-enhancing gases have been emitted in meaningful quantities, the temperature rise has been less and, as we shall see, hardly indicative of a greenhouse catastrophe.

The visual impact of Hansen's curve is striking, for it appears to rise dramatically at the end. In fact, the last point is so warm that it is almost off the chart; we can appreciate the biasing effect of this point by obscuring it with a thumb. That last point does carry a notation saying that it represents the temperature departure from normal for the first five months of 1988, but it does not say what that departure means.

Hansen knew that any variation in a sample of monthly temperatures is by definition greater than that observed in yearly data. For example, two-thirds of all the annual mean temperatures in the United States are within 1°C (1.8°F) of the long-term annual mean. The variance of monthly temperatures, particularly in the cold half of the year, is approximately three times as great. Consequently, if an aggregate of monthly temperatures is plotted against annual values, the monthly temperatures are likely to appear extreme. And when all of the 1988 data finally came in, the annual value, while warm, did not look at all unusual when compared with the rest of the data for the warm 1980s.

Why did Hansen not note the problem of inflated variance in his testimony? It is difficult to believe that he was unaware of it and equally unlikely that many, if any, of the onlooking members of Congress and their staffs knew that they had been less than fully informed. As to the fourth estate, how many reporters and television correspondents know the science of statistics?

Most scientists who work on or follow the issue do. And that is why it was so disheartening to many and so embarrassing to the profession when the Union of Concerned Scientists sent an update of the same distortion to tens of thousands of citizens, asking that they be concerned about global warming and contribute a few dollars more.

The die was cast, and the contributions rolled in. Hansen's testimony resulted in a vast percentage of the American public believing the climate apocalypse was at hand; it helped to create the Popular Vision of gloom and doom. And the same people who telephoned CNN's 900 poll at personal expense will surely vote, something no official can ignore. Thus did a simple distortion (perhaps inadvertent), covered dramatically by a press that did not know better or would not say if it did, reinforce a popular perception that global ecocide is at hand.

Hansen received a great deal of media attention as a result of his June 23, 1988, testimony. After a few appearances on the late night shows and then on Dan Rather's prime-time show (where he left the public with the definite impression that 1988 mega-hurricane Gilbert was a by-product of global warming), he finally made it to the home of the best Sunday buffet in Washington, "This Week with David Brinkley." He brought along as a prop a pair of dice on which two-thirds of the dots were painted to indicate "hot" and the other dots implied otherwise. His purpose was to demonstrate that an enhanced greenhouse effect does not mean that every year will be hot, but rather that the probability increases that a given year will be above normal. That ploy seemed somewhat amusing since he had taken no great pains to emphasize to Congress the statistical incongruity of comparing a five-month sample to annual data.

At any rate, one scientist—who cannot be named here, but the story is all around the profession—was sufficiently irate that he went down to the local five-and-dime store for two sets of fuzzy dice—one red and one orange. He ripped off all the little felt dots, substituting the words "oranges" on the orange dice and "apples" on the red ones. He then broke the sets apart and retied them so that each set consisted of an "apple" die and an "orange" one and sent them along to Jim. Rumor has it that the terse accompanying note read, "Next time you compare five-month data to annual averages in front of Congress, maybe you ought to roll out these babies."

Who Pays?

There is no doubt that the amount of CO_2 in the atmosphere is increasing and that it is not the only greenhouse gas whose

19

atmospheric concentration we are increasing. Doubtless, the major source of CO_2 is the combustion of fossil fuel, and coal produces the most CO_2 per unit of released energy. Three nations—the United States, the former Soviet Union, and continental China—contain the vast majority of world stocks of coal; Australia is the largest exporter. In the United States, coal provides the primary source of external capital for many states, especially West Virginia, Montana, Wyoming, and Virginia.

If we assume that the apocalyptic visions of climate change are true, a 60 to 80 percent reduction in greenhouse emissions is required to prevent ecocide. Because of its relatively high production of CO_2 per unit of energy, the obvious first target for CO_2 reductions is coal. Legislation submitted in 1990 to the House of Representatives required industries that produce very modest emissions of CO_2 to "offset" those emissions with some program, such as tree planting, that would remove an equivalent amount from the atmosphere.

Rapidly growing trees sequester carbon dioxide in the form of carbohydrates; the irony is that mature trees and forests (such as those that are home to the famous Spotted Owl) are not growing very much and do not take up any net CO_2. A serious attempt to mitigate CO_2 with trees should promote the cutting of mature forests and prevent the subsequent re-release of the CO_2 by preserving that wood for a long time (i.e., turning it into houses).

Forestation programs usually require the planting of trees on something other than the emitter's land. Cheap land that will lend itself to that type of long-term sequestration will become increasingly scarce and unproductive. Because coal produces more CO_2 per unit of energy than any of the other major fossil fuels, the coal industry or utilities that use coal will have to disproportionately incur the cost, while fossil fuel sources that produce relatively less CO_2 (such as natural gas) will become more attractive, although gas, too, produces considerable CO_2. Substitution of natural gas for coal (which produces about 50 percent of U.S. power) in all electricity production would not come even close to the emission reductions necessary to forestall the Popular Vision. Nonetheless, Louisiana would prosper at the expense of Virginia, and futile attempts to fight global warming would become a source of income redistribution from the haves to the have-nots.

20

According to the World Bank, the ratio of CO_2 emissions to gross national product of the United States is very efficient in comparison to that of other nations of similar geographic size and transportation needs (see Table 2.1). Nonetheless, most modest energy companies would have to engage in expensive programs that would eventually entail substantial consumer cost. For example, a relatively small North Carolina utility has estimated the cost of complying with proposed legislation at $2 billion.

The greenhouse-active nitrogen oxides are thought to originate mainly from the use of high-energy nitrogen fertilizers and some subsequent atmospheric reactions. Those fertilizers are the chemicals that drive the American agricultural machine, a principal source of foreign exchange.

Table 2.1
CARBON DIOXIDE EMISSIONS PER UNIT OF ECONOMIC
ACTIVITY FOR GEOGRAPHICALLY LARGE NATIONS

Nation	Emissions Mt CO_2/yr X 10^6	GNP ($bn/yr)	CO_2/ GNP
Mainland China	2,236.2	372.3	6.01[a]
South Africa	284.2	79.0	3.60
India	600.6	237.9	2.52
Mexico	306.9	176.7	1.74
Soviet Union	3,982.0	2,659.5	1.50[a]
Canada	437.8	435.9	1.00
United States	4,804.1	4,880.1	0.98
Australia	241.3	246.0	0.98
Brazil	202.4	323.6	0.63
European Community[b]	2,424.0	4,182.3	0.58

SOURCE: World Bank, Washington, D.C.
NOTE: We often hear that the European Community is almost "twice as efficient" as the United States with respect to the production of CO_2. In terms of raw numbers, that is obviously true. But when that efficiency is adjusted for the relative geographic area (which is a measure of transportation needs per capita), the European Community emits almost three times as much CO_2 as the United States.
[a]Probable underestimate because of currency overvaluation.
[b]Germany, United Kingdom, Spain, Italy, and France.

U.S. corn yields averaged around 30 bushels per acre in the 1930s. Thanks to genetic improvements such as development of hybrid corn, enhanced efficiencies in tillage as tractor replaced horse, and increasing use of nitrogen fertilizers, yields now average over 100 bushels per acre (a threefold increase in the past 50 years).

Very little of that increase would have taken place without the great rise in application rates of nitrogen fertilizer. That increase is responsible for the transformation of the United States from a modest exporter of grain in the early part of the century into the Saudi Arabia of agriculture (see Figure 2.5).

Even so, most of the primary productivity of American agriculture is not consumed directly by humans, in part because most grains are seriously or completely lacking in one or more of the

Figure 2.5
CORN YIELDS IN THE SOUTHEASTERN UNITED STATES, 1960–85

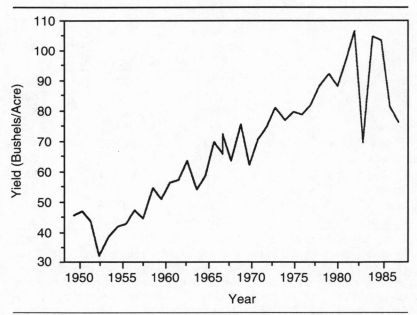

NOTE: One of the common myths about agriculture is that crop yields are declining because of environmental degradation. This plot of southeastern corn yields is typical of most of the world's granary nations.

amino acids that we humans cannot synthesize directly but nevertheless require as constituents of our protein. So, most American grain is cycled through animals and consumed either in the form of bovine dairy products or directly as meat, with beef forming a substantial percentage of that mix. Ruminants, such as cows, produce large quantities of methane in their digestive tracts, and bovine flatulence is thought, along with rice paddy agriculture, to be a major source of the observed increase in that gas.

Thus, the United States' most abundant domestic source of fossil energy, coal, and our prolific font of foreign exchange, agriculture, are the principal sources for the gases that are responsible for three-quarters of the potential for global warming. Doubtless, any attempt to reduce emissions of those gases by 60 to 80 percent would dramatically affect our entire culture. Throw in an attempt to proscribe rice paddy agriculture (the cultural centerpiece of hundreds of millions in the Orient), and the prevention of global warming becomes by far the greatest centrally planned social engineering experiment in history.

If our policy is going to be commensurate with our science, the probability of an apocalyptic climatic change must be very high indeed.

23

3. The Intergovernmental Panel on Climate Change

In 1990 the United Nations published its *Scientific Assessment of Climate Change*, authored by the Intergovernmental Panel on Climate Change (IPCC) and commonly referred to as the IPCC report (UN 1990). That report is extensively cited by environmentalists, lawmakers, and interested citizens as providing the consensual basis for dramatic reductions in greenhouse gas emissions.

Approximately 200 scientists, bureaucrats, and administrators were invited to contribute various chapters on agriculture, ecosystems, and climate history, but the document was really crafted by a small number of lead authors (usually one to three people per subject area) who were able, with input from the 200, to tailor each chapter as they best saw fit. An additional team of approximately 100 reviewers critiqued the original document, and again the lead authors had final say about admission of a reviewer's comments.

That document is often held up as the "consensus of scientists," but it is much more the consensus of a very carefully chosen group of lead authors. Later in this book I will document the profound lack of consensus that exists among even the community of scientists and reviewers who worked on the IPCC document. As both a contributor (to the 1992 update) and a reviewer (UN 1990), I can personally attest to that.

The overall IPCC document is several hundred pages long and prefaced by a brief Policymakers Summary, which is the section that has been read the most, and it is doubtful that many officials who speak so glibly about the IPCC consensus ever got beyond that summary. The summary was prepared by the IPCC Working Group 1, chaired by Sir John Houghton of the United Kingdom Meteorological Office.

Speaking to a 1991 meeting of climatologists in Asheville, North Carolina, Dr. Houghton's immediate understudy, the very capable

Christopher Folland, explained the difficulties that were encountered in producing the summary of the overall report. The entire document gives a fairly comprehensive description of the climate problem, and, although it is not always candid about the limitations of, say, the models we use to relate climate change to forest change, it is about as evenhanded a document as one could get from the competing egos that produced it.

Dr. Folland related that the problem in creating a summary was the somewhat equivocal nature of the entire report. Of what use to policymakers, he asked, would a document be if punctuated by repeated cautions and "on the other hand" statements. So, Folland stated, the group made a conscious decision to produce the Policymakers Summary that did not equivocate as much as the overall report.

The Policymakers Summary, which serves as a useful touchstone for analysis of the climate change problem, is preceded by an Executive Summary in which the following points are made.

I. Certainties

    A. The greenhouse effect is real and greenhouse gases "already" keep the earth warmer than it would otherwise be.

    B. Manmade emissions will increase the concentration of greenhouse gases, including water vapor, which will lead to a warming of the earth's surface.

II. Calculations Made with Confidence

    A. Reductions in greenhouse gas emissions will take from decades to centuries to become fully effective.

    B. Stabilization of greenhouse gases at present levels would require a reduction of manmade emissions of more than 60 percent.

III. Predictions

    A. At the current growth rates in greenhouse gases, the rate of increase in global mean temperature during the next century will be about 0.3°C (0.5°F) per decade. Temperatures will increase by 1.0°C (1.8°F) in the next three decades.

    B. Land surfaces will warm more rapidly than the ocean, and high northern latitudes will warm more than the global mean in winter.

    C. The temperature increase in southern Europe and central North America will be higher than the global mean.

    D. Sea level will rise 20 centimeters (8 inches) by 2030 and 65 centimeters (26 inches) by the end of the next century.

IV. "Our Judgment Is That . . ."

    A. Global mean surface temperatures have increased 0.3°C to 0.6°C (0.5°C to 1.1°F) in the past 100 years.

    B. The five warmest years were in the 1980s.

    C. "The size of that warming is broadly consistent with predictions of climate models. . . ."

    D. Episodes of high temperatures will become more frequent in the future simply because mean temperature will increase.

    E. Many ecosystems that have been disadvantaged by climate change will be unable to migrate fast enough and will be vulnerable to extensive damage caused by exceptional events such as drought and fire.

The Policymakers Summary then expands on the points made in the Executive Summary. Relevant aspects of the Policymakers Summary are detailed below.

V. How Might Humans Change the Climate?

    A. An increase in greenhouse gases will raise temperatures and change other aspects of climate, particularly precipitation and evaporation.

    B. Changes in greenhouse gases were part (but not all) of the reason for the pulsing of the ice ages.

VI. Human Enhancement of the Greenhouse Effect. Because of the combined effect of all of the greenhouse-enhancing gases, the effective increase in CO_2 is 150 percent of its level in 1750.

VII. What Will Be the Patterns of Climate Change in the Next Three Decades?

    A. The land will warm faster than the oceans; the Northern Hemisphere will warm faster than the southern.

B. In the high latitudes of the Northern Hemisphere, winter warming will average 75 percent more than global mean warming and summer warming will be "substantially smaller" than the global mean.

C. Where we currently have reliable climatic records, the "best estimates" (assuming that there is no change in the current rate of emission increases) of regional climate change for the next three decades are:

1. Central North America (i.e., the agricultural heart of the United States) will experience
    a. An average warming of 3.0°C (5.4°F) in winter
    b. An average warming of 2.5°C (4.5°F) in summer
    c. Soil moisture decreases in summer averaging around 18 percent.

2. Southern Europe will experience
    a. An average warming of 2.0°C (3.6°F) in winter
    b. An average warming of 2.5°C (4.5°F) in summer
    c. Soil moisture decreases in summer averaging around 20 percent.

3. Australia will experience
    a. An average warming of 2.0°C (3.6°F) in winter
    b. An average warming of 1.5°C (2.7°F) in summer.

VIII. Other Expected Climate Changes

A. The number of "very hot days" or frosty nights will change substantially. The number of days with a minimum threshold amount of soil moisture (for viability of a certain crop, for example) will be sensitive even to changes in average precipitation and evaporation.

B. Our forecast models give no clear indication of whether or not hurricanes and midlatitude (winter) storms will increase or decrease in frequency or intensity.

IX. Confidence in These Climate Predictions

A. Because warming "will lead to an increase rather than a decrease of the natural greenhouse gas abundance, . . . climate change is likely to be greater than the estimates we have given."

B. During the next century, climate change from increasing greenhouse gases is likely to be far more important than any mitigating effect, such as other industrial emissions that might offset warming.

C. "We have substantial confidence that models can predict at least the broad-scale features of climate change."

X. Have These Changes Already Begun?

A. The overall temperature rise has been broadly similar in both hemispheres.

B. Confidence in the observed warming of surface temperatures has been increased by their similarity to recent satellite measurements of midtropospheric temperatures.

C. "The size of the warming over the last century is broadly consistent with the theoretical predictions of climate models, but it is also of the same magnitude as natural climate variability. If the sole cause of the observed warming were the manmade greenhouse effect, then the implied climate sensitivity would be near the lower end of the range inferred from the models."

At various points in the text that follows, I will refer to some of those outline statements and discuss whether or not they are supported by observations and data. Readers may wish to refer to the list more frequently.

4. Computer Models: Calculating the Apocalypse

Our understanding of the factors that are involved in climate and climatic change has improved considerably since Arrhenius made the first enhanced greenhouse calculations in 1896, but scientists are still grappling with the problem of simply simulating, from first physical principles, the regional and seasonal climates of the planet. Although it is a modestly straightforward calculation to estimate the global mean temperature by considering incoming and outgoing energy (called the "energy balance method"), such calculations can reduce the earth to a single point, instead of specifying the required regional climates of its diverse surface that result from the fact that the temperature is not the same everywhere. Those calculations, therefore, give a correct but meaningless answer, for unless regional climate—such as that of the corn and soybean belt or the high Arctic—can be reliably simulated and projected, we simply cannot predict with any confidence the ecological implications of a change in the concentration of greenhouse gases. In other words, within certain broad limits, *how much* the mean temperature of the earth as a whole changes may be irrelevant. What matters is *how* the climate changes. The "how" questions include matters of timing, seasonality, spatial distribution of changes, and even whether or not changes are equally distributed between day and night.

The most detailed simulations of climate—and those that attempt to answer the "how" questions—are known collectively as General Circulation Models (GCMs). Simple in principle but very complicated in operation, GCMs attempt to objectively calculate both regional and seasonal climate and how that climate should change if the greenhouse effect were enhanced.

Like the work of so many scientists (this author included) concerned with the problem of global warming, the GCMs were not intended to become objects of public scrutiny and policy debate.

Rather, they were designed as teaching and research tools for advanced study of the atmosphere. They were not originally intended to "forecast" the future, and most principal investigators spend much more time talking about the limitations of GCMs than they do touting their predictions.

It was not until 1975—right in the middle of the period when the Popular Vision of climate was that of an impending ice age—that the first influential GCM–carbon dioxide paper appeared. And I believe it is fair to say that its authors, Syukuro Manabe and Richard Wetherald, had no intention that their paper be used to launch the grandest experiment in the central planning of energy ever attempted in the industrial democracies.

That paper and a subsequent modification in 1980 projected that doubling CO_2 would lead to a global warming of approximately 4.0°C (7.2°F). Neither paper is easy reading, and I suspect that many who seized on either as a vehicle for promoting the global warming issue never read either from beginning to end. If they had, they would have found that the calculations were based on assumptions about the basic physics and characteristics of the atmosphere that are known to be wrong. For example, the amount of solar radiation that low clouds (the most common type in the earth's atmosphere) reflect was lowered by 10 percent from its known value. That lowering has the effect of increasing the amount of radiation warming the atmosphere by approximately 3 percent, which is roughly equivalent to moving the earth about 2 million miles closer to the sun. A 2 percent increase in the earth's overall reflectivity sends so much of the sun's energy back out to space that it cancels the warming expected from a doubling of CO_2.

The lowering was necessary because the computer, when left to its own devices, simulated the pre-greenhouse-enhancement climate at around 5.0°C (9.0°F) colder than the present temperature. That calculation would have meant the mean temperature was approximately the same as at the height of the last ice age and would have implied that glaciers should still be a few thousand feet thick over Chicago.

That altered "parameterization," as it is called, is hardly an act of chicanery, for the authors duly noted it in their scientific communication. And that alteration was not the only one. The authors discovered that the GCM was misprogrammed to melt the north polar ice cap at what is roughly today's temperature. If that were true, the earth would have no ice cap. An earth without a north polar ice cap would warm more because what is now a great reflective region would become a dark, energy-absorbing surface. The difference between white and dark can be appreciated by taking a barefoot walk on a sunny day first over white concrete and then over blacktop.

The authors then corrected the error by keeping the ice cap intact at the nonenhanced temperatures, and essentially the same warming was produced. Early GCMs, whose oceans were effectively only around 100 feet deep and had no currents, were bent on producing a warm planet no matter what. By 1986 the Geophysical Fluid Dynamics Laboratory (GFDL) model, which had been used as the benchmark for a comprehensive Department of Energy review of global warming, had bumped up the amount of solar radiation that hits the top of the atmosphere by 100 watts per square meter, or 7 percent above its known value. If we were the several million miles closer to the sun that that error would require, we would be living on a much warmer planet.

Those anomalies did not stop Washington and may even have goaded policymakers on. Disregarding the model's apparent proclivity to produce apocalyptic results, regardless of the physics involved, the federal government began holding extensive meetings on what to do about global warming. A 1979 meeting in Annapolis, sponsored by the Department of Energy, was replete with sociological and psychological impact panels, a sure sign of approval inside Washington's Beltway. Climatologist Reid Bryson of the University of Wisconsin told me that participants were informed that there would be no questioning of Manabe and Wetherald's GCM, which was the scientific centerpiece.

By the mid-1980s five major GCMs had acquired scientific prominence, although several others had produced somewhat different results. The colloquial names, which serve as useful referents, are given here. First, the Princeton model (also called the GFDL model) was developed primarily by Manabe at the U.S. Department of Commerce's Geophysical Fluid Dynamics Laboratory in Princeton, New Jersey. Second, the NASA model (also called the GISS model) is managed by James Hansen of the Goddard Institute for Space Studies. Third, the NCAR model is mainly a result of the work of

Warren Washington at the National Science Foundation–sponsored National Center for Atmospheric Research in Boulder, Colorado. Fourth, the OSU model, named for Oregon State University, was developed primarily by the RAND Corporation in Santa Monica, California, and is now mainly associated with Michael Schlesinger of the University of Illinois (Urbana-Champaign). Schlesinger is the member of the modeling community who has been the most vocal about the extraordinary uncertainties, ambiguities, and unknowns associated with that type of work. Under somewhat mysterious circumstances he walked out of a meeting of the UN Intergovernmental Panel on Climate Change (the "consensus" of climatologists?) in Great Britain. Finally, the UKMO model is investigated by J. F. B. Mitchell of the United Kingdom Meteorological Office in Bracknell.

The primitive mid-1980s versions of the models, which were the ones most featured in congressional hearings on global warming, have been continually refined by the addition of, say, oceans that actually have currents or thunderstorms that are less than 288,000 square miles in minimum extent (in reality, even the largest midlatitude summer thunderstorm complexes are not much bigger than 50,000 square miles, and most are less than 360). But the early versions of the models predicted, on average, that the mean surface temperature would warm 4.2°C (7.6°F) as a result of doubling the initial concentration of CO_2. In the high latitudes of a Northern Hemisphere winter, the projected changes averaged around 8°C (14.4°F), and in some areas they were nearly 20°C (36°F).

In spite (or perhaps because) of such wild speculations, problems similar to those that plagued the GFDL model also plague others. In a 1992 volume of the scientific journal *Geophysical Research Letters*, John Walsh of the University of Illinois wrote that currently used GCMs routinely have errors of 5°C to 10°C (9°F to 18°F) in the Arctic winter even before increases in CO_2 are considered. One of the persistent rumors in climate modeling, though it has never been publicly pronounced, is that the current NASA model, when asked to simulate an ice age on its own, promptly melts the polar ice cap with very little CO_2 in its air. You can imagine what it does at today's concentrations. The original UKMO runs—like those of other 1980-era GCMs that have been fuel for the apocalyptic vision—also produced very unrealistic results unless the ocean temperatures were a priori "prescribed" (i.e., fed in as answers).

34

Return to I.C of the IPCC outline in the previous chapter and note that "we have substantial confidence that models can predict at least the broad-scale features of climate change."

We obviously cannot reliably simulate the current climate from first principles, so the question of "how much do we expect the climate to change from what it was?" becomes impossible to answer. In recognition of that problem, a very clever switch is made: model output is not presented as the expected change from, say, the current climate or even the global climate around 1900. Rather, it is presented as the difference between two model runs: one with $1 \times CO_2$ and the other with $2 \times CO_2$. The "forecast" is actually one model, complete with the systematic errors described above (and many, many more), subtracted from an equally flawed "background" model.

Are the problems still with us? On September 23, 1987, James Hansen was asked specifically by the Environmental Protection Agency in Raleigh, North Carolina, what the greatest problem was with GCMs, and he replied that it was their inability to properly estimate regional climate.

Therefore, the predictions of regional climate change noted in the IPCC outline are unreliable. Why were they even made?

Nine months to the day later, Hansen gave his famous testimony to Congress about a strong "cause and effect" relationship between current climate and human alterations of the atmosphere. His words caused 70 percent of the respondents to a CNN poll to blame the (regional) U.S. drought on the greenhouse effect, something Hansen had just told the EPA that his models could not reliably predict.

In a table that accompanies a 1991 article in *Nature*, Wei-Chung Wang demonstrates that the "background" (i.e., $1 \times CO_2$) simulation projected by a new generation of the same NCAR model discussed above indicates that the earth should now be about as cold as it was at the end of the last ice age (Figure 4.1).

Or consider the most advanced American GCM, the coupled GFDL atmosphere-ocean model published in the August 1991 *Journal of Climate*. It assumes that the energy coming from the sun is some 20 watts per square meter less than it really is so it can achieve a realistic $1 \times CO_2$ earth. That error would not be so bad if the results were intended only for academic consumption. But the

Figure 4.1
NCAR Model's Projected Change in Mean Temperature (°C) for December–February

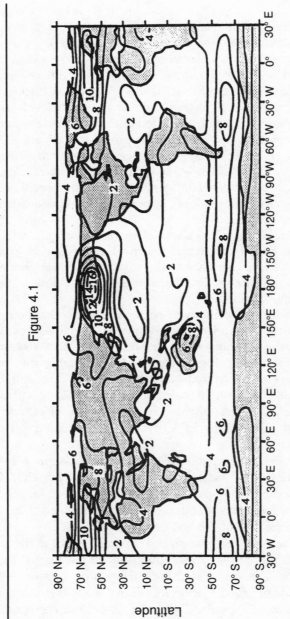

Figure 4.1

Longitude

Latitude

NOTE: December–February mean temperatures (°C) for doubled CO_2 in the NCAR model are subtracted from those simulated with "background" CO_2. This model is actually the "coolest" of the GCMs of the mid-1980s.

output for doubled CO_2 is being used as a vehicle for making public policy.

Normally, around half of the sun's radiation actually goes to warm the atmosphere, so the effect of unrealistically decreasing the sun's output by 20 watts translates to about 10 watts at the planet's surface. The problem is that enhancing the greenhouse effect by doubling CO_2 is equivalent to increasing radiation at the earth's surface by about 2.5 watts per square meter. Consequently, the "adjustment" factor used to get the right temperature before an increase in CO_2 is factored in is several times greater than the change that one would expect a change in CO_2 to cause.

Imagine that today's weather forecast for the corn belt was, say, 10°C (18°F) too hot, but the forecaster knew it and arbitrarily subtracted the error. That is what occurs in climate models. In the weather forecast situation, all subsequent computer products that would be made public, ranging from the probability of severe thunderstorms to the agricultural weather forecast for crop spraying, would appear ever so reasonable but would be driven by a model that was producing large known errors of agricultural consequence. If you knew an error was lurking in the weather forecast, would you remake your daily agricultural plan as a result?

If you knew there were very similar and, unfortunately, very real problems (as opposed to the weather-forecasting one I just made up) in the GCMs, would you, in the words of Worldwatch, use those models as a scientific basis for "remak[ing] the world's way of life"?

> There will be no sudden change. There will be those who don't agree, but as soon as the man in the street notices, it won't matter. If the model is correct, the increased frequency of drought will be evident in the 1990's, in the early 1990's if there is no large volcanic eruption.
>
> —NASA scientist James E. Hansen
> quoted by Richard Kerr in
> *Science*, June 2, 1989

Taken as an aggregate, the GCMs of the mid-1980s predicted a net warming of 4.2°C (7.6°F) after the earth adjusts to a doubling of the background CO_2 concentration. As we have already seen, we are more than halfway there because of the additional effect of increases in the non-CO_2 greenhouse enhancers. At current rates of increase, the CO_2 concentration will effectively double by 2030 or 2050.

The response of the mean temperature to changes in CO_2 is not linear; rather, it is a logarithmic function that implies that the first half of an increase in CO_2 should create much more warming than the second. But because our effective CO_2 concentration is increasing exponentially—the mirror image of the logarithm—our temperature should be undergoing a rapid linear rise, as shown in Figure 4.2. Thus, if something were not holding back the warming, the

Figure 4.2
CLIMATE RESPONSE TO A CHANGE IN CO_2

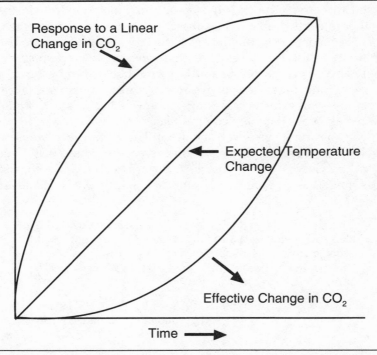

NOTE: This schematic diagram shows the climate response to a change in CO_2. Because the response to a linear increase is logarithmic (top curve) and the actual increase has been exponential (bottom curve), we would expect a roughly linear change in temperature to be observed if something (such as the oceans) were not holding back the warming.

global mean temperature, according to the models, should be more than 2.1°C (3.8°F), or one-half of 4.2°C (7.6°F) higher than that observed before major emissions of CO_2-enhancing compounds.

But the GCMs all indicate that, for a doubling of CO_2 above the background, warming is maximized in polar winter, particularly in the Northern Hemisphere, in some areas of which the temperature increase would range from 12°C (22°F), according to the NASA model, to 18°C (32°F), according to the GFDL model. For comparison, the current annual range in mean monthly (difference between January and July) temperature in Atlanta, Georgia, is 36°F. Obviously, profound changes in the high latitudes should already be apparent.

Again, if we assume nothing holds back the warming, the January temperature in the high latitudes of the Northern Hemisphere should now be several degrees warmer than it was. However, that is a somewhat unfair restriction inasmuch as much of the accelerated warming is expected to take place after some recession of high-latitude ice fields. Rather, there should now be a considerable warming slightly south of those regions—around latitudes 60°–70° N, especially in winter.

Tropical regions are not projected to warm as much as the rest of the globe because tremendous changes in energy are required to warm the surrounding saturated atmosphere. As an example, consider the opposite, deserts. The daily solar cycle from sunrise to sunset is sufficient to warm the temperature some 33°C (60°F) in the Sahara. Just a few miles to the west, over the Atlantic Ocean, the same change in solar energy warms only the lowest layers of the atmosphere some 7°C (13°F). The minuscule changes in surface warming energy for doubled CO_2 (noted earlier as 2.5 watts per square meter, or less than most flashlights), therefore, can exert only a tiny warming in humid environments. In fact, the effect of all the moisture in oceanic regions is to effectively cap the earth's temperature. With few exceptions, ocean surface temperatures rarely exceed 31°C (88°F). Further, according to calculations by V. Ramanathan (Ramanathan et al. 1989) of Scripps Institute of Oceanography, if the tropical ocean temperature rose only two more degrees, the consequent development of explosive thunderstorms would terminate almost all additional warming.

Warming is also not projected to be equally distributed between the hemispheres. Differences in surface characteristics, including the amount of land and the tenacity of ice caps (the Antarctic ice cap is, on average, around 500 times thicker than the thin band that sits atop the Arctic Ocean), lead to calculations of less net, or much slower, warming in the Southern Hemisphere. It follows that the Northern Hemisphere, as a whole, should have, to date, warmed more than the Southern Hemisphere and more than the expected global average. Table 4.1 shows what warming should have occurred according to the GCMs of the mid-1980s if the oceans were not holding back warming and under the more realistic assumption that the ocean must retard some of the warming. According to the IPCC, if all warming during the 20th century is assumed to be a result of greenhouse enhancement, the net warming for a doubling of CO_2 would be only about 1.3°C (2.3°F); however, the average of GCM-projected warming is 4.2°C (7.6°F).

Table 4.1

EXPECTED AND OBSERVED CURRENT WARMING FROM THE GREENHOUSE ENHANCEMENT IN °C (AGGREGATE OF MID-1980s GCMs)

Region	Assuming No Lag	With Ocean Lag	Observed (Past 50 Years)
Globe	2.1°	1.2°	0.25°
Southern Hemisphere	1.2°	0.7°	0.4°ª
Northern Hemisphere	3.0°	1.7°	0.1°
Tropics	0.8°	0.5°	0.4°
High latitudes, NH	4.0°	2.3°	−1.0°ª

NOTE: It is clear that observed warming has been far beneath the projections. The specifications come from a variety of sources but are quite representative of most parameters of climatic change that have been used as the basis for current policy recommendations. I assume an expected warming to date of 2.1°C (3.8°F) with no lags, because that is one-half of the net warming expected for a doubling of CO_2 in the aggregate of the mid-1980s climate models. In a 1987 article in *Climate Monitor*, Tom Wigley of the University of East Anglia calculated that, even assuming a very strong oceanic lag, the global mean temperature by 1985 should have been some 1.2°C (2.2°F) above the background. I therefore use the scaling factor of .57 (1.2/2.1) to very liberally estimate the lag in warming. The Northern/Southern Hemisphere differential is abstracted from a GCM published by Manabe and Wetherald in the 1989 *Journal of Physical Oceanography*. Because of the geography and the great humidity of the tropics, warming there is projected to be only 40 percent of the global average, and the equilibrium high-latitude Northern Hemisphere warming assumes its typical value of 8.0°C (14.4°F). "Observed" temperatures are linear trends during the past 50 years in the global records of Phil Jones, also of the University of East Anglia, and Tom Wigley. Those are the records most frequently cited in the scientific literature and are described in detail in succeeding sections of this book.

[a]The Jones and Wigley Southern Hemisphere records do not include data from Antarctica. John Sansom (1989) of the New Zealand Meteorological Service has analyzed those records. If his analysis is factored into the Jones and Wigley history, the observed Southern Hemisphere warming becomes 0.35°C (0.63°F).

5. Taking the Earth's Temperature

In the parlance of science, the GCMs describe testable hypotheses about what might happen to the climate in an enhanced greenhouse. To test those hypotheses, we must determine what has happened as greenhouse gas concentrations have increased. If the observed temperatures and warming support the GCM projections, then they would seem to be an adequate basis for public policy. If they do not support the GCMs, then those models are inappropriate bases for policy.

Unfortunately, taking the world's (or the hemisphere's) mean temperature from ground-based thermometers is not easy. We cannot simply take readings from all the weather stations in the world and average them because of several complicating factors.

Further, the first large area thermometer networks were put in place in the late 19th century at the close of a 200-year cold period known as the "little ice age." The June 1992 issue of the American Geophysical Union's newsletter *EOS* carried an article demonstrating that other long-term records, which can be used to estimate temperature, show that a similar warming occurred after another cold period a few hundred years ago.

Instruments and their locations have changed over time. In the early 20th century, most "official" thermometers were located on the north walls of buildings; now most reside in little white boxes known officially (and quaintly) in the United States as "cotton region shelters." Those boxes, in turn, are rapidly being replaced by smaller ceramic "beehives" that are part of newer automated transmitting stations. The person who knows more than anyone in the world about long-term temperature records, Tom Karl of the U.S. Department of Commerce, thinks that the simple change from north walls of buildings to cotton region shelters may have artificially increased recorded temperatures a few tenths of a degree. Little white boxes heat up more in the daytime than big buildings and transmit that heat to the recording thermometers.

The problem with temperature measurement can be seen almost every night on local weather programs. Most TV meteorologists now run networks of "volunteers" who telephone in readings, and their readings are presented on the screen to show how warm it got at grandma's house. Within each TV station's range there is usually also some "official" measurement point—often an airport—and a reading point at or near the center of a decent-sized city.

Those three sets of data all give different temperatures. The readings from the center of the city, usually measured in cotton region shelters, are almost always warmer than the rural airport temperatures (also measured in shelters) because bricks, buildings, and pavements retain more heat and prevent normal ventilation.

Washington, D.C., provides a very good example of that phenomenon. Washington's temperatures have been skyrocketing since the 1920s, and the warming is concurrent with the growth of the federal government. In fact, the mean January temperature measured at the urbanized Washington National Airport is now warmer than that measured at the rural Leander McCormick observatory, some 100 miles to the south. Although there are many uncertainties in the study of climatic change, one thing we are sure of is that Washington, D.C., did not move 100 miles south. Instead, the infrastructure that goes with economic activity (in Washington it might be the waste heat from all that money changing hands) warmed it up.

The high temperature readings from volunteer observers, who usually have either beehive or liquid-in-glass thermometers that they attempt to shield from the sun, are normally several degrees warmer than either of the other two, because it is very difficult, especially in summer, to find a thermometer exposure that is not hit by the sun at some time of day.

Which reading is "correct" or truly indicative of regional temperature under undisturbed conditions? Most probably it is the rural airport temperature. Unfortunately, commerce and population have a way of gravitating toward those airports, and after two or three decades, those records, too, begin to show artificial warming that has nothing to do with the greenhouse effect.

There are other very insidious biases in the longest standing records. Most originated at 19th-century points of commerce, which means they were near sources of waterpower (read "rivers"). Those sites, in which cold air pools at night, then show artificial local warming years later as the city grows up around the station. Thus, they are initially shielded from a true climatic warming, and then they exaggerate one that may not have occurred. Although we know quite a bit about the effect of cities themselves on the temperature, the magnitude of the combined effects has never been calculated.

Or consider land-use changes when forested land is turned into farmland, noting what happened in the U.S. corn belt in the summer of 1988. And often, when land gets too enthusiastically farmed or populated, as in Sahelian Africa or Northern Mexico, farmland turns into desert, which changes the regional climate until pressures on the land are relaxed.

How bad are those effects? Tom Karl has developed the most reliable regional climate record by scrupulously noting movements of stations and population changes, but his work is currently applicable only in the United States where station records are (fairly) easily accessible. He found that statistically significant artificial warming of temperatures begins to appear in towns with populations as small as 2,500. After sifting through the 16,000 official temperature stations that are in the U.S. Department of Commerce's national network, Karl retained fewer than 500 in what he calls the Historical Climate Network (HCN). Although that sample is easily representative of an area the size of the conterminous United States, the dismissal of 97 percent of the original stations demonstrates how pervasive the urban effect can be in biasing our temperature records.

After constructing the HCN, Karl requested from James Hansen of NASA the data that he used from the same region when he constructed the climate record that set Congress afire in his June 1988 testimony. When the two data sets were compared, a remarkable discrepancy arose: Hansen's record showed a statistically significant warming trend of 0.44°C (0.8°F) that simply does not appear in the carefully constructed HCN record. Indeed, the HCN shows no warming whatsoever in the past 60 years. The results of the comparison were published in the March 1989 *Bulletin of the American Meteorological Society* (see Figure 5.1). Yet not one newspaper story appeared, even though the error—if extended globally (and there was no reason to believe it could not be)—would mean that the primary basis for Hansen's testimony was fiction.

Figure 5.1
DISCREPANCY BETWEEN NASA'S RECORD AND KARL'S HCN

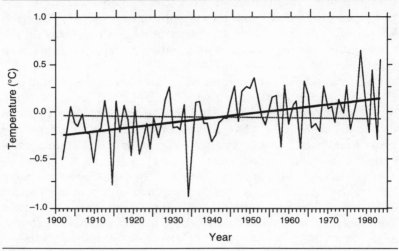

NOTE: Differences between the NASA record and Tom Karl's records of the United States, as presented in the *Bulletin of the American Meteorological Society*. It turns out that the wrong data were shipped by NASA to Karl for comparison.

In the same article Karl also compared the U.S. portion of the record from East Anglia University (the previously mentioned Jones and Wigley history) to the HCN and found a smaller error, on the order of 0.10°C to 0.15°C (0.2°F to 0.3°F). According to the Jones and Wigley record, the overall global warming trend over the past 100 years (or since reliable records began) is 0.45°C (0.81°F).

The discrepancies begat a great deal of scientific scurrying. Jones and Wigley seemed quite willing to accept such a small bias in their record, even though if extended globally it would lower the warming of the past 100 years to between 0.30°C and 0.35°C (0.54°F and 0.63°F). Is that amount of warming really "broadly consistent with predictions of climate models," as the IPCC outline says?

Hansen and his coworkers simply could not believe their data were that bad. They were not, and soon his people discovered the source of the problem. When they programmed the computer to present the U.S. data so they could be sent to Tom Karl for comparison, an error was made and a lot of data from the eastern Pacific

Ocean and Mexico were sent instead. The problem was probably caused by a computer instruction that specified where and when to extract data.

That unfortunate incident emphasizes the problems associated with making statements with a "high degree of confidence" about a "cause and effect" relation between current temperatures and human greenhouse enhancement. The data for, say, the 1980s may be significantly different from those for some earlier period at the 99 percent statistical confidence level (meaning there is only a 1 percent probability that the difference in the mean value of two sets of data is due to pure chance), but are the data themselves correct?

Another problem results when recording stations are moved. In the spring of 1988 Hansen and coworker Sergei Lebedeff published a paper demonstrating that 1987 was a very warm year globally, and every major newspaper in the nation covered the story. The data came from the ground-based thermometer network that was used to generate the temperature history Hansen presented to Congress on June 23, 1988. Some of those data suffer from the effects of cities' growing up around the thermometers, but an attempt was made to remove part of that problem. Nonetheless, some stations kept showing excessively warm temperatures; the most glaring was an island station, Santa Helena, in the southern Atlantic Ocean. Seasonal departures from the previous 30-year "normal" were consistently in the range of or in excess of 2°C (4°F), which is phenomenal for an island sitting in the tropical Atlantic. The station is very important because there are no others for hundreds of miles, and it is used to determine the temperature of a large region between Africa and South America.

A call to the British Meteorological Office, which runs the station, revealed that in the 1970s (during the end of the 1951–80 period that was used as a reference for normal values), the station had been moved some 600 feet down a mountainside. No wonder all the temperatures are now warmer than normal.

The global thermometer-based histories that I use in subsequent sections are those of Jones and Wigley. Even so, the correlation between those records and the highly accurate satellite records that are described below is not particularly great. The statistical correspondence between the two records is only about one-third, meaning that two-thirds of the variability that shows up in the satellite record does not appear in Jones and Wigley's records.

47

Further, a peculiar discrepancy between the Jones and Wigley land-based records and oceanic ship records was corrected in a very controversial fashion. Generally speaking, records of oceanic temperatures, which are called sea surface temperature (SST) records, show less warming than land records. That difference is as it should be, because the ocean obviously warms more slowly than the land, as was noted earlier. The records were reconciled by arbitrarily raising some ocean temperatures to be equal to land temperatures. That change induces additional warming in combined ocean-atmosphere records. The argument could just as easily have been made that all the coastal stations with a potential for artificial warming (i.e., those whose populations that exceed 2,500) should have had their temperatures adjusted down to the SST value.

> Trend analysis of long-term records of land and ocean temperatures and sea level are qualitatively consistent with the climate changes projected by modeling studies.
>
> —*Detecting the Climatic Effects*
> *of Increasing Carbon Dioxide,*
> U.S. Department of Energy (1986),
> page xxvi

> Even if the warming [of the past 100 years] has been due to CO_2, its magnitude is almost a factor of 2 less than expected from those model results in which the time lag induced by the oceans is only a few decades [see points IV.C and X.C in the IPCC outline].
>
> —Ibid., page 171

According to a 1991 *Nature* article by GCM modeler Michael Schlesinger, the most likely lag is on the order of a few decades.

Now the usual procedure in the greenhouse discussion is to present the Jones and Wigley global temperature history and to argue that the observed warming of the past 100 years appears to be equal to or slightly below the lowest limits suggested by mid-1980s GCMs. Environmentalists will assert that the global curve is within the projections, while scientists who work with the climate data recognize a dearth of warming.

In either case, reference to the global temperature history fails to further the debate because GCMs project considerable differences in behavior between and within hemispheres. If the apocalyptic vision of global warming is to be verified, the earth must warm up in a fashion that is consistent with apocalyptic models. A mere change in the mean annual temperature of the planet, within certain broad limits, says nothing about that prospect.

The Southern Hemisphere Temperature Record

In the Southern Hemisphere 90 percent of the surface is water, and the highest 20 degrees of latitude are covered with ice that averages thousands of feet thick. Liquid water requires a great deal more energy to raise its temperature a given amount than does an equivalent land surface. Further, the snow and ice fields of Antarctica and vicinity, because of their inherent brightness, reflect more than three-quarters of the incoming solar radiation. For comparison, the earth as a whole absorbs about 75 percent of the solar radiation that reaches its surface.

It follows that temperature variations should be much less in the southern half of the globe than they are in the north. The increase in radiation that is trapped because of greenhouse gases should currently show much less of an effect than it does in the Northern Hemisphere because that radiation is warming water, not land.

The Jones and Wigley record certainly shows some warming during the 20th century (Figure 5.2). The net change since 1900 is about 0.6°C (1.1°F).

Unfortunately for catastrophists, it is difficult to ascribe that change solely to an altered greenhouse. The vertical double bar in Figure 5.2 marks the beginning of the period in which the important greenhouse gases began to be emitted in large quantities. That period (since World War II) coincides with the industrial development of much of the world; a virtual population explosion of flatulent bovines; and the industrial use of CFCs, which are very effective greenhouse absorbers. In fact, around two-thirds of the enhanced greenhouse effect has been induced since 1950, and a very slowly increasing fraction of one-third occurred during the 150 or so years between the Industrial Revolution and World War I.

Clearly, at least half the warming took place before the major emission of the greenhouse enhancers began. If the warming of the early 20th century is attributable to the greenhouse, then today's temperatures should be much warmer than shown.

50

Figure 5.2
Jones and Wigley's Record of Temperatures in the Southern Hemisphere, 1860–1990

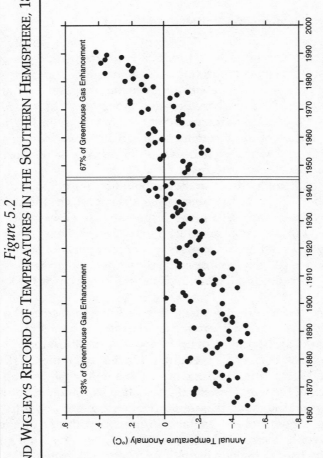

NOTE: Two-thirds of the greenhouse enhancement has taken place to the right of the double bar. It is quite difficult to attribute much of the temperature rise of the early 20th century to an altered greenhouse. Further, this record does not contain John Sansom's Antarctic data that appeared in *Journal of Climate* and showed no net change in temperature since reliable records began in 1957.

Nonetheless, the record for the Southern Hemisphere does seem consistent with some greenhouse enhancement. There is an apparent warming of 0.3°C (0.5°F) since 1950 (with 0.2°C or 0.4°F in the brief period since 1979). Although the figures themselves are not large, it is generally conceded that the Southern Hemisphere should warm up least and slowest.

It is worth noting, though, that the record does not include any Antarctic data. In 1989 New Zealand's John Sansom published that history, which extends back only to 1957, in the *Journal of Climate*. When observed warming is averaged across the entire land mass, there is no net warming. Consequently, the recent warming in the Jones and Wigley record must be tempered somewhat. Using a scaling factor that is appropriate for the relative size of Antarctica, the net warming in the Southern Hemisphere should be reduced by approximately 0.05°C (0.09°F), resulting in a quarter of a degree of warming since 1950.

Comparison to the Satellite Data

In 1979 NASA launched the first of a series of satellites that could measure the temperature of the lower atmosphere with an accuracy of ± 0.01°C (0.02°F). The measurements are based on the vibration of oxygen molecules, which changes with temperature. Satellite coverage is also universal, whereas the ground-based temperature network is confined primarily to land areas, which constitute only 30 percent of the planetary surface.

The companion ocean temperatures in the ground-based networks, such as Jones and Wigley's, have serious problems because of the temperature of ships, which tends to be warmer than that of the water, and the methods by which the measurements are taken. They used to be taken by hauling aboard a canvas bucket of seawater, which evaporated and cooled in the wind. Later, thermometers were installed in the intake tubes for seawater used for engine cooling. Especially in cold environments, it is likely that the mechanical warming of the ship's surface biases those temperatures upward. Thus, the transition from bucket to intake temperatures introduced some artificial warming into the record. Attempts have been made to remove that error, but there is no way of knowing how many other systematic errors the correction process may have introduced. Finally, as was noted earlier, sea surface temperatures

51

were arbitrarily adjusted upward to agree with nearby land temperatures, which is a highly debatable proposition.

The major problem with the satellite record is that the sensing
units were not in orbit until 1979, so the record is only 14 years long.
Nonetheless, it coincides with a rapid warm-up of approximately
0.25°C (0.45°F) in the Jones and Wigley record of the Southern
Hemisphere.

The satellite record is also amenable to comparison, on a monthly
basis, with the various land-based records. The tool to measure
correspondence is a figure called the "explained variance" between
sets of temperatures. If the satellite and land temperatures corres-
ponded perfectly, the explained variance between the two would
be 100 percent. In that type of comparison, an explained variance
of less than 50 percent is poor.

As expected, the explained variance between the satellite record
and Tom Karl's HCN—the one that is most carefully monitored for
urban warming—is the greatest, at 86 percent. Unfortunately, that
record, which shows no net change in temperature for the past 60
years, applies only to the conterminous United States. The hemispheric records of both Jones and Wigley and NASA show very
poor correspondence. The explained variances between the satellite
records and the former are 42 percent and 20 percent (Northern
and Southern Hemispheres, respectively), and they are 46 percent
and 34 percent for the NASA record. Yet the IPCC outline states,
"Confidence in the observed warming of surface temperatures has
been increased by their similarity to recent satellite measurements."

There is a remarkable discrepancy between the satellite record
and Jones and Wigley's Southern Hemisphere temperatures: the
satellite record shows no warming since 1979 (see Figure 5.3), and
the disparity between the satellite and the Jones and Wigley records
is growing.

Plotted in Figure 5.4 are departures from the Jones and Wigley
mean for 1979–90 minus the same information for the satellite data;
a trend line is run through the figures. If there is no discrepancy,
or if it is random from year to year, the trend should be flat. If there
is a systematic increase in warming in the Jones and Wigley record
that does not show up in the satellite data, the trend should point
upward, which it does with a vengeance: almost 0.4°C (0.7°F) per
decade.

Figure 5.3
SATELLITE RECORDS OF TEMPERATURES IN THE SOUTHERN HEMISPHERE

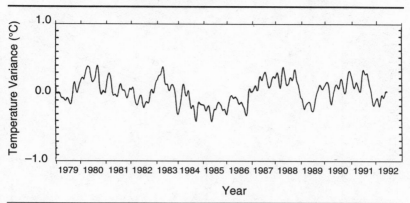

NOTE: The satellite-sensed temperature records of Spencer and Christy (1990) show no significant warming in the Southern Hemisphere.

Is the Jones and Wigley temperature record for the past 12 years more a history of how rapidly population is growing in the Southern Hemisphere than it is one of the true temperature of the region? Australian W. S. Hughes recently sent data to climatologist Robert Balling detailing Australis's climate history. He examined the 13 long-term stations that make up the climate record there. Six are in state capitals that are rapidly growing cities in an arid environment. In a similar situation in the United States, Phoenix shows an artificial warming of several degrees, which dwarfs the well-documented spurious warming of Washington, D.C., by almost an order of magnitude.[1]

The National Oceanic and Atmospheric Administration (NOAA) has recently constructed a "HCN-like" record, going back to 1950,

---

1. The possible alternative explanation to urban contamination—that somehow the satellite data are in error in the Southern Hemisphere—seems ludicrous, except for the fact that we know that stratospheric temperatures (the region of the atmosphere between approximately 7 and 50 miles above the surface) have cooled dramatically there in the past 10 years. Are satellite data somehow including some of the stratospheric temperatures while the satellite attempts to monitor the lower atmosphere? In a careful reworking of their data, in which they measured different layers of atmosphere (as presented at the National Climatic Data Center in Asheville, North Carolina, on August 13, 1991), Spencer and Christy unassailably argued no.

Figure 5.4
SOUTHERN HEMISPHERE TEMPERATURES FROM SATELLITE RECORDS COMPARED TO LAND RECORDS

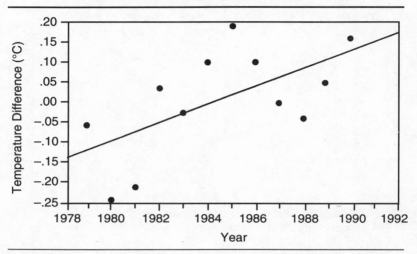

NOTE: Satellite temperature departures from the mean over the Southern Hemisphere subtracted from the same of Jones and Wigley since 1979—the period of concurrent records. The trend line, objectively calculated by standard mathematical methods, reveals a growing warm bias of nearly 0.4°C (0.7°F) since 1979 in the Jones and Wigley record.

of seasonal temperatures for each hemisphere. It is not as good as the true U.S. HCN, but it is probably more reliable than the Jones and Wigley history, although its limited duration—only 40 years—does not allow us to say much about 100-year trends. But the behavior of the Southern Hemisphere in the NOAA record between 1979 and 1990 (when it is concurrent with the satellite data) may well confirm that the fairly large warming in the Jones and Wigley record is spurious, for the NOAA record also shows no warming at all during the period. Nonetheless, the NOAA record does show some prior warming (from 1950 to 1979) south of the equator. Thus, the bottom line appears to be that the Southern Hemisphere's warming is real for much of the past half century, but that urban contamination has drastically compromised the data from the past decade or so in the Jones and Wigley history.

The Northern Hemisphere Temperature Record

While the modest Southern Hemisphere warming may be explainable in part by of the predominance of water and ice, the implication is that the Northern Hemisphere, which contains most of the world's land, should have warmed much more and much faster.

It did not (as shown in Figure 5.5). In fact, in the standard statistical sense, which is the 95 percent confidence level, there is no net warming in the past 50 years, although the record is extremely noisy. However, if the confidence interval is relaxed to 10 percent, there is a warming of 0.15°C (0.27°F) over the past 50 years. In the overall record, the most prominent feature of the past century is a rapid warm-up of 0.5°C (0.9°F) that took place between 1915 and 1930. Because that increase occurred so early, it could hardly have had anything to do with the enhanced greenhouse. Thus, the "natural" variability—how climate changes with or without human influence—of the Northern Hemisphere must be on the order of at least 0.4°C (0.7°F), which suggests that the greenhouse signal will be very difficult to find in our temperature record.

A 1986 paper by Hugh W. Ellsaesser from Lawrence Livermore Laboratory demonstrated how difficult it is to even notice those types of climatic change. He showed that if one adds 0.25°C (0.45°F) to all the Northern Hemisphere temperatures recorded before 1917 and subtracts the same from all readings taken after 1921, there is no net change in the entire record from the time it begins in the 19th century to the early 1980s. The implication is that there was a very brief (five-year) warming, starting during Woodrow Wilson's presidency, that explains the entire trend for 100 years. What is most remarkable is that no one noticed it until 66 years later. That warming was much more rapid than any projected by GCM climate models for the next 50 years.

Comparison to the Satellite Data

Northern Hemisphere satellite data (Figure 5.6), like those for the Southern Hemisphere, show no warming since the platforms were launched in 1979. Moreover, the very warm years of the 1980s, which are so evident in the land-based record, simply do not appear in the satellite readings.

Figure 5.5

JONES AND WIGLEY'S RECORD OF TEMPERATURES IN THE NORTHERN HEMISPHERE

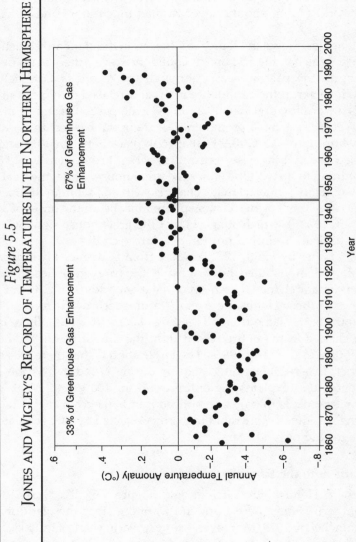

NOTE: The Northern Hemisphere temperature history of Jones and Wigley shows almost all of its warming *before* the major greenhouse enhancement.

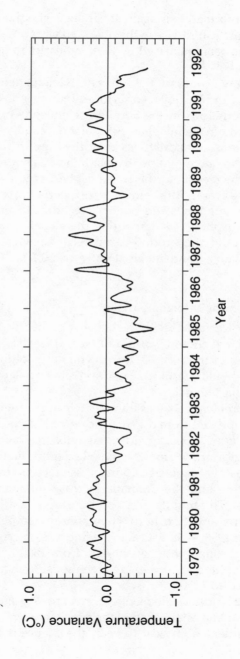

Figure 5.6

SATELLITE-SENSED TEMPERATURE RECORDS OF SPENCER AND CHRISTY FOR THE NORTHERN HEMISPHERE

That information is again at variance with the IPCC outline statement that "confidence in the observed warming of surface temperatures has been increased by their similarity to recent satellite measurements."

When we compare the Northern Hemisphere satellite data and the Jones and Wigley records to see if there is an increasing discrepancy (as there is in the Southern Hemisphere), we find none. In the Northern Hemisphere, population changes, on a percentage basis, are dramatically less than they are in the Southern Hemisphere, so one might not expect to find as strong an urban-related bias in the recent Northern Hemisphere record. But what happens when we average the hemispheres together to compare global satellite data with Jones and Wigley's record? The biasing of the Southern Hemisphere is so great that the global difference between the satellite and the ground-based records, when objectively determined with a trend line, is slightly over 0.3°C (0.5°F) for the past 12 years.

Hemispheric Intercomparison

Thus, we are left with the peculiarity that the land hemisphere—the one that should warm up first and fastest—shows virtually no warming in the past 50 years, while the Southern Hemisphere—which should warm up least and slowest—gives a more "greenhouse-like" signal.

The problem becomes immediately apparent in Figure 5.7, which I have updated from a 1990 article by Sherwood Idso in *Theoretical and Applied Climatology*. Idso has published approximately 300 climatological articles in the refereed scientific literature. The top chart is the history of global carbon emissions from the U.S. Department of Energy, and the dramatic increase following World War II is obvious. The middle and bottom graphs are Jones and Wigley's Northern and Southern Hemisphere temperatures from 1880 through 1990. A statistical trend line was objectively calculated and placed through both sets of data from 1880 to 1950, which marks the beginning of the great greenhouse enhancement, and then extended to 1990.

In the Northern Hemisphere, *every reading after 1950* falls beneath the trend line established in the previous 70 years. If readings were independent from year to year, the chance that that would occur

Figure 5.7
GLOBAL CARBON EMISSIONS COMPARED TO
HEMISPHERIC DATA

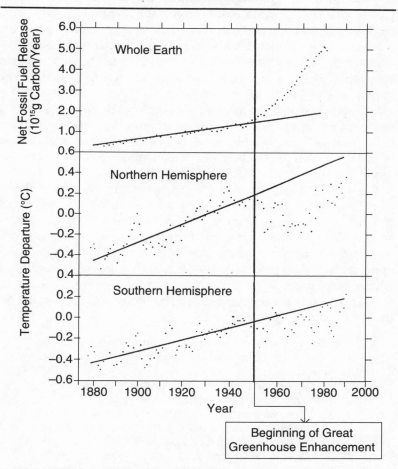

NOTE: In this figure, which is updated from a paper by Sherwood Idso, the top plot is global carbon emissions, which rise dramatically after World War II. The middle and bottom plots are Jones and Wigley's Northern and Southern Hemisphere data, respectively. A linear trend line has been placed through the data from the beginning of the record until emissions began their dramatic increase. It is apparent that the hemisphere that should show that most warming *stopped* warming at the rate that was established *before* the major greenhouse enhancement.

randomly (assuming that there had been no physical change in the previously determined system) for 40 consecutive years is 1 in 2^{40}, or 1 in 1.09 trillion. In reality, readings are not independent from year to year (i.e., a warm year is more likely to follow a warm year than is a cold one), but the degree of association is not that great. If we make the indefensibly liberal assumption that the interdependence of the data reduces the number of independent observations by a factor of 5, the chance of that behavior occurring is still very remote at 1 in 256.

More perplexing is the fact that the Southern Hemisphere continues to warm, although it, too, runs a bit below the trend established there from 1880 to 1950. Has there been some fundamental change in the atmosphere of the Northern Hemisphere that has not occurred in the southern half of the planet?

The problem becomes even more apparent when Southern Hemisphere temperatures since 1950 are subtracted from those of the Northern Hemisphere (Figure 5.8). Because the Northern Hemisphere should be the first and fastest to warm, that difference should be increasing significantly. As shown in Figure 5.8, the opposite has taken place, and the statistical confidence in the trend is very, very high.

That behavior contravenes several statements in the outline of the IPCC Policymakers Summary:

> III.B Land surfaces will warm more rapidly than the ocean, and high northern latitudes more than the global mean in winter.
>
> IV.C The size (of the observed) warming is broadly consistent with predictions of climate models.
>
> VII.A The land will warm faster than the oceans; the Northern Hemisphere will warm faster than the southern.
>
> IX.C We have substantial confidence that models can predict at least the broad-scale features of climate change.
>
> IX.A Because warming "will lead to an increase rather than a decrease of, the natural greenhouse gas abundance, . . . climate change is likely to be greater than the estimates we have given."

Thus, a tremendous inconsistency has arisen. *Every* GCM projects that Southern Hemisphere warming should be slower than that for the Northern Hemisphere, and yet the opposite is taking

Figure 5.8
DIFFERENCE BETWEEN NORTHERN AND SOUTHERN HEMISPHERE TEMPERATURES, 1950–90

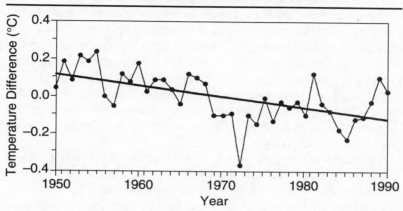

NOTE: The hemispheric differences, northern minus southern, since 1950 should be increasing. Instead, there is a highly significant *decline*, which demonstrates that the wrong half of the planet has been warming up. "If you're a believer," Tom Wigley of the University of East Anglia told *Science* magazine on February 7, 1992, "you can see the Northern Hemisphere warming more slowly than models predict." This plot underscores that no faith is required.

place. The Northern Hemisphere has most of the world's land, most of the world's industry and greenhouse emissions, a very thin ice cap, ephemeral snow cover, and a surface dramatically altered by settlement and agriculture—all of which should enhance warming. The Southern Hemisphere is almost all water at its surface, has very few people and little industry, and has permanent snow and ice cover thousands of feet thick—all of which should retard warming. And yet it is the Northern Hemisphere that shows no net warming in the past 55 years.

GCMs are great research and teaching tools, but is there any defense for using the GCMs of the mid-1980s as a basis for, in Lester Brown's words, "a wholesale reordering, a fundamental restructuring of the world's economy," when they have the world turned upside down?

Rapid Warming: The Popular Vision

One of the arguments commonly heard at cocktail parties is that "all scientists agree the greenhouse effect is real" followed by the observation that "the six warmest years in the record are all in the past decade." Although it would be surprising if they were not, because the urban biasing effect has to be greatest at the end of the record, the fact is that their warmth is far beneath the projections implied by the climate models of the mid-1980s. Close inspection of the Northern Hemisphere record—which should show the most rapid warming—reveals that the warm years of the 1980s are less than *two-tenths* of a degree (C) above values frequently observed elsewhere in the record.

An interesting response can usually be elicited by asking, "How warm were those years in the 1980s?" The common retort is that they were 2°C to 4°C (3.6°F to 7.2°F) warmer than before and that the ice caps are melting. In fact, most apocalyptics are incredulous when they learn that there has been so little warming and that the Greenland ice sheet—the largest glacier in the Northern Hemisphere—is growing.

Given that minuscule warming, we must ask, "Is outline statement IV.B—that the five warmest years were in the 1980s—a deliberate attempt to mislead?" The amount of warming in those years of the 1980s is so minuscule that it is doubtful any human being alive could sense it. However, thanks to weekly page-three bludgeonings in the newspapers, everyone feels hot, and the policy express is in throttle-notch nine. Imagine what would happen if what happened 70 years ago were to occur again (see Table 5.1).

The consequences of that *natural* variation, if it occurred again, say, beginning in 1994, would be profound, probably involving the

Table 5.1
RELATIVE TEMPERATURES, 1917–21

Year	Relative Temperature (°C)
1917	0.00°
1918	0.20°
1919	0.26°
1920	0.39°
1921	0.45°

White House, Congress, and considerable costs. Unfortunately, something very similar may happen as the cooling effect of Mt. Pinatubo's 1991 eruption relaxes in 1993.

The Vision of *Science*

"Global Warming Continues in 1989," trumpeted the headline of a "news" piece by editor Richard Kerr in the February 2, 1990, issue of *Science*. The fact is that there was no global warming trend in the Jones and Wigley record for the decade of the 1980s. There is, however, a warming in the Southern Hemisphere (where it should not be), but the Northern Hemisphere data are so noisy that they swamp the trend from the other half of the planet.

What did occur was a rapid global warming from 1977 through 1980, followed by no change. Any global trend in the 1980s is statistically indistinguishable from zero. It is peculiar that global warming received much more attention years after the temperature changed than it did while the change was occurring. Is the implication that there was very little noticeable impact as it happened? Consider that the warming from 1917 through 1921 was not noticed until Hugh Ellsaesser of Lawrence Livermore National Laboratory "discovered" it 66 years later.

All of those things were noted in a letter to *Science* dated February 12, 1990, but *Science* declined to publish it. The reason? "That's not news. Everyone knows there was no warming in the past decade."

> If the global warming situation is analyzed using the customary standards of scientific inquiry, one must conclude there has been much more hype than solid fact.
>
> —Philip Abelson
> editorial comment in *Science*, March 30, 1990

High-Latitude Temperatures

If there is any consistency among the GCMs that fired up Congress, it is that they all predict dramatic warming for an effective doubling of atmospheric CO_2 during north polar winter. It could even be worse: environmental scientist Daniel Lashof of the Natural Resources Defense Council has warned of a further dramatic warming because heating those regions would unfreeze enough permafrost to rapidly release methane, another potent greenhouse gas.

There is one interesting catch: if warming tends to concentrate in a high-latitude winter, then ipso facto it occurs predominantly during the night or when the sun is very near the horizon. At the poles, night is 183 days (one-half of a year) long; poleward of 70° the sun is below or near the horizon for months on end. That is obviously not the growing season, so it is impossible to stress what few plants there are (buried under feet of snow) with that type of warming. It is also difficult to melt the ice cap with a modest temperature rise when the mean temperature averages around −40°C (−40°F).

For the observed increases in greenhouse gases, the NCAR GCM projects that high latitudes in North America should have warmed some 2°C to 4°C (3.6°F to 7.2°F) by now, and the NASA model has been quite adamant for years that the high latitudes of the Northern Hemisphere should be very warm.

The search for the great Arctic warming has been on for some time. The first place to go is obviously to the temperature histories, and they behave in a fashion that is the opposite of what has been forecast. For example, the Jones and Wigley record north of 55° (Figure 5.9) shows a net cooling for the past half century and a dramatic warming before that—when the greenhouse enhancement was minimal.

Compare that information with two statements in the outline of the IPCC Policymakers Summary:

> The warming will average 75 percent greater than the global mean warming in the high latitudes of Northern Hemisphere winter and "substantially smaller" than the global mean in high latitude summer. . . .
> We have substantial confidence that models can predict at least the broad-scale features of climate change.

It is hard to understand why the work of James Lachenbruch and his colleagues at the U.S. Geological Survey in Menlo Park, California, has received so much public attention. Their paper, which was published in a November 1986 issue of *Science*, described temperature profiles in holes drilled into the Alaskan permafrost and noted that there had been a warming of 2°C to 4°C (3.6°F to 7.2°F) "somewhere during the past 100 years." Unfortunately, that period includes the rapid warming of the early 20th century that can hardly have anything to do with an enhanced greenhouse effect.

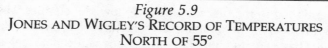

Figure 5.9
JONES AND WIGLEY'S RECORD OF TEMPERATURES
NORTH OF 55°

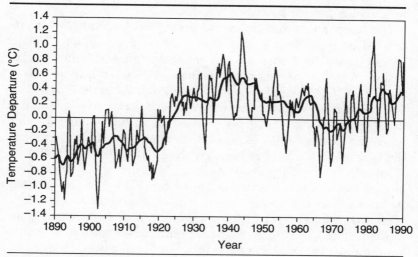

NOTE: These data show warming before the major increases in trace gases and a net cooling in the past half century.

Although Lachenbruch's paper prompted a number of newspaper stories, a scientific communication criticizing any greenhouse implications of his work was sent to *Science,* and Lachenbruch replied that he had no intention of ascribing the apparent warming to the enhanced greenhouse.

At any rate, it is evident from the Jones and Wigley record that there was a dramatic warming of the Arctic sometime during the past 100 years—it just happened to take place long before the trace gas enhancement became important. In fact, both the temperature records and Lachenbruch's work demonstrate that the climate of the Arctic is highly variable—with the obvious "natural" variability somewhere in the range of 2°C to 4°C (3.6°F to 7.2°F).

What is perhaps more disturbing is the degree to which so much of the scientific community is simply unaware of the observed temperature records. When Lachenbruch's work received another airing at the U.S. Geological Survey's annual Global Change meeting in Reston, Virginia, in March 1991, the presenter seemed genuinely taken aback when the Jones and Wigley temperature history was mentioned in the discussion that followed his paper. His surprise underscores the amazing degree to which environmental advocates have created a feeling of consensus, even among scientists, that is not supported by the data.

The temperature history of the polar North Atlantic, as documented by J. T. Rogers in 1989 at NOAA's 13th annual Climate Diagnostics Workshop, is at even greater variance with the Popular Vision. Winter temperatures—the ones that should be going up— show a 3°C (5.4°F) decline from 1920 through the early 1980s. That finding was not very surprising to the scientists who live with the data, but it prompted a page-three story in the *Washington Post* when Jay Zwally of NASA and others documented in *Science* that the Greenland ice cap has been growing rather than shrinking.

John Sansom of the New Zealand Meteorological Office has also analyzed Antarctic records. Most notable is the record from Amundsen-Scott, the south polar station. There, the sun is below the horizon for six months of polar night, and there is surely no "urban effect." During the night a scientist might hope to see some warming, considering the large increase in effective CO_2 that has taken place to date. Nonetheless, the accompanying chart of his data (Figure 5.10) shows no warming whatsoever. While the greenhouse effect "must be real," it is therefore also possible that the layers of the atmosphere very near the surface during polar night may be shielded from the brunt of much of the dramatic warming.

The Siberian Express

Arrhenius's calculations, made back in 1896 and described at the beginning of this book, indicated that an enhanced greenhouse would result mainly in a warming of winter and night temperatures. The new GCMs do not differ much, except that some are so crude that, to save on computer time, they do not have day and night. It is always a sunny day.

It follows that if winter and night temperatures are those most affected by an enhanced greenhouse, the weather systems that form during winter night would be those that should change the most. Those are the cold high-pressure systems of polar origin that set most of the winter temperature records east of the Rockies. Though most of them originate in northern Canada or over the high Arctic, a few actually are detachments of the airmass that sits over

Figure 5.10
SANSOM'S RECORD OF ANTARCTIC TEMPERATURES

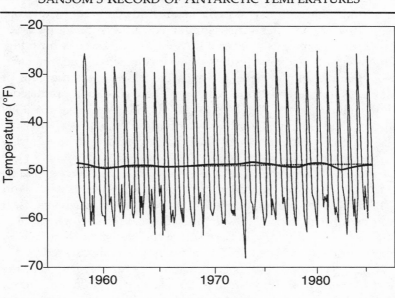

NOTE: Reliable records of Antarctic temperatures really did not begin until the International Geophysical Year of 1957, but records at the South Pole (where there is only one day/night cycle per year) surely show no warming. John Sansom's (1989) data appeared in the *Journal of Climate*.

Siberia most of the winter and causes Siberians to record some of the Northern Hemisphere's coldest temperatures. The cold airmasses that come screaming down into the southeastern corner of North America (i.e., through the eastern half of the United States) have come to be known as the Siberian Express.

Much of the 19th century saw a viable citrus industry that stretched from Grand Isle, Louisiana, all along the Gulf and Atlantic coastal regions and northward to approximately Charleston, South Carolina. February 1899 changed things.

Early that month, the jet stream contorted into an orientation that stretched the normally frigid airmass, which sits in Siberia all winter, a bit closer to the North Pole and the high Arctic of North

America. The pile of cold air was sufficiently deep in the high latitudes that barometric pressure records were set that stood for most of the 20th century. When a fast-moving storm drew the cold air southward, temperatures plummeted in the U.S. southeast to values never previously recorded. At 7 a.m. on Lincoln's birthday, Tampa reported moderate snow (a few inches actually accumulated over much of Florida) and a west wind blowing off the Gulf of Mexico at an average 24 miles an hour. That condition produced a wind chill factor of $-22°C$ ($-8°F$), and the Siberian Express of 1899 so damaged the more northerly citrus groves that economically viable citrus production has since been confined to the Florida peninsula.

From 1899 through 1918 there were 18 major Arctic outbreaks, many of which originated in Siberia, into the southeast. And 1917–18 was the worst period, with a total of eight such events. It is not surprising that 1917 was the nadir that forms the beginning of the very rapid five-year warm-up of the Northern Hemisphere, referred to earlier, that has only recently been "discovered."

A history of those events is shown in Figure 5.11. Only 5 of the next 40 years (from 1919 through 1958) saw major Arctic outbreaks, and during the entire period there were only 10 such events, 6 of which were clustered in 1934–36. That means, not counting that brief period, that there were only 4 such events in the entire 40 years. Citrus crept back northward through Florida.

That state of affairs changed in 1959, and no one knows why. In the face of the dramatic enhancements of the greenhouse effect that really began around 1950, the Siberian Express returned after a virtual absence of 40 years. Since 1959 there have been 24 big outbreaks—we have returned to the frequency that characterized the early 20th century—and some have been whoppers. Many of the barometric records that were set in 1899—indicating a cold airmass of record depth—were broken in 1989. Planes were grounded in northwestern North America because altimeters can be adjusted only as high as a surface pressure of 31.00 inches. Any reading beyond that is something their manufacturers obviously consider beyond the scope of reasonable probability. Several studies indicate that the Arctic outbreak of January 1985 (which forced President Reagan's second inauguration indoors) was the coldest airmass to reach southeastern North America in the 20th century.

Figure 5.11
ARCTIC OUTBREAKS IN VIRGINIA, 1897–1989

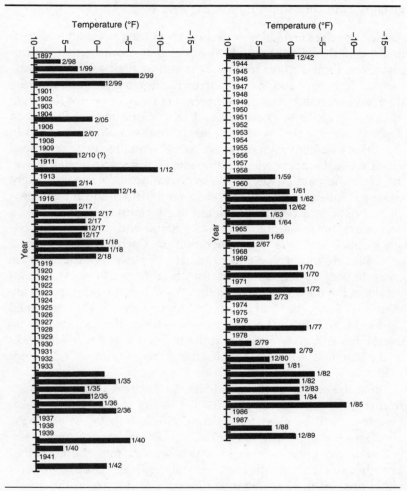

NOTE: Years without Arctic outbreaks are shown as blank. The temperatures shown are the mean values observed over Virginia on the coldest night of each event. In the face of the substantial increase in greenhouse gases, the frequency of years that report these events since 1959 is almost an order of magnitude greater than it was in the prior 40 years. Compare this with statement VIII.A in the outline of the IPCC Executive Summary: The number of "very hot days" or frosty nights will change substantially.

Since 1959 citrus culture has been forced ever further south on the Florida peninsula, and citrus trees' northern limit of economic viability is further south than it has ever been.

Arctic airmasses form almost exclusively during polar night or twi-light—and therefore should be the ones that show dramatic warming. Even though their frequency is still as high as it was in the early 20th century and they still break the low-temperature records, they have warmed some at their point of origin. Laurence Kalkstein of the University of Delaware and the Environmental Protection Agency recently surveyed the high latitudes of North America to see if there was any sign of what should be an obviously enhanced greenhouse. He acknowledges that there has been no overall warming over the past 50 years; the evidence actually favors a slight cooling. However, Kalkstein went a step further and broke down all the data into objectively determined "airmass type" days, a procedure that is mathematically formidable and was computationally difficult until the new generation of supercomputers became available. He found that approximately 15 percent of the days are characterized by airmasses that can be referred to as Arctic outbreaks in the making. He also found that when only those days are examined, it appears that the coldest Arctic airmasses have warmed up by 2°C (3.6°F) in the past 40 years—even though there is no overall net annual warming and the frequency of Arctic outbreaks in the southeastern United States has increased dramatically since 1959.

Thus, the erroneously named Siberian Express appears to warm from −40°C to −38°C (−40° to 36.4°F) as it passes over Alaska. Few tears will be shed from Fargo, North Dakota, where those airmasses freeze people, to Florida, where they freeze orange trees, if that is the only airmass in which we can find the enhanced greenhouse effect. At present, the only animals that live under it either are sleeping, are wrapped in acres of fur, are in the ocean, or, if they are *Homo sapiens*, are dressed up like the Michelin tire man.

Are the Ice Caps Melting?

Much was made recently of news reports that the north polar ice margin has shrunk by 2 percent since 1979 and that North American snow cover has shrunk to its lowest value since then. Those observations were made from satellites, and some of the orbiters were in place long before 1979.

In 1980 Don Wiesnet and Michael Matson of NOAA published an article in *Environmental Data Information Service Reports* describing satellite measurements of snow cover since records began in 1966. As can be seen from the accompanying illustration (Figure 5.12), maximum snow cover in the Northern Hemisphere rose by approximately 18 percent between 1966 and the late 1970s.

Starting a record at its high point and using subsequent trends to promote the vision of climate apocalypse is a common technique, and it is unfortunate that the records described in the two recent reports just discussed begin in the late 1970s. But the polar story, including high-latitude temperatures, the Siberian Express, and the record of snow cover, reveals how little of that vision is based on observed data. And it is in the polar regions that the most dramatic changes should already be obvious.

Further, much of the furor over the melting of ice in Greenland or Antarctica, which is required to raise seas to apocalyptic levels, is scientifically groundless within any reasonable statute of limitations. A warming of the air temperature by even several degrees over those areas allows the moisture content of the atmosphere to increase, resulting in more precipitation. But the temperature is so far below freezing that all that moisture falls as snow. It may seem paradoxical, but the way to make the world's two major ice fields grow is to warm things up a bit.

Thus, still more ocean water would be sequestered as ice in the two gigantic sheets, making it even less likely that sea level would rise drastically. We, therefore, have to ask the basis for and the purpose of the parade of Public Broadcasting Service television programs, such as "After the Warming," which show the oceans rising sufficiently to climb up the side of the Washington Monument. A similar publicity stunt, sponsored by the Environmental Defense Fund and the Smithsonian Institution, is now traveling from museum to museum. Why is it necessary to propagandize our children with such nonscience?

Perplexed? Disbelieving that global warming could increase the high-latitude glaciers and snow fields? Read from the November 1991 journal *Geology*. On the basis of meticulous studies of marine

71

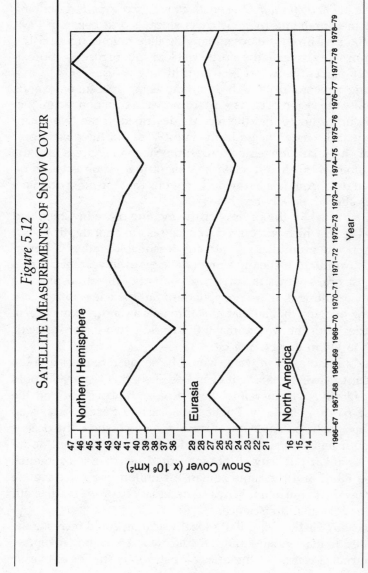

Figure 5.12
SATELLITE MEASUREMENTS OF SNOW COVER

NOTE: In 1991 we were regaled with reports that the polar ice volume had diminished by 2 percent since 1979. In fact, satellite-measured snow cover in the Northern Hemisphere had risen by 18 percent since records began in 1966 and the maximum that was observed in the late 1970s.

sediments around Antarctica dating back to the "climatic optimum" (that's what pre-1980 climatology texts—published before the global warming apocalypse became fashionable—called the period between 4,000 and 7,000 years ago that was around 2.0°C [3.6°F] warmer than today), Eugene Domack from Hamilton College and his coworkers found that "temperature-accumulation relations for the Antarctic indicate that warming will cause a significant increase in accumulation [of snow and ice] rather than ablation [recession]."

What about the Arctic? Gifford Miller of the University of Colorado and Anne de Vernal from the University of Quebec wrote in the January 16, 1992, issue of *Nature:*

> We find that the geological data support the idea that greenhouse warming, which is expected to be most pronounced in the Arctic and in the winter months, coupled with decreasing summer insolation (increasing cloudiness) may lead to more snow deposition than melting at high northern latitudes, and thus to ice-sheet growth.

Do not be surprised if the nouvelle apocalypse, revised version for the year 2000, will be that global warming may cause sea level to fall, exposing Bangladesh to wrenching cultural changes. For that reason we will need to, in Lester Brown's words, undertake a "wholesale reordering, a fundamental restructuring of the world economy." It does not matter whether sea level goes up or down, whether the temperature goes up or down or remains the same: whatever happens, some people just seem to think that the world needs more central economic planning.

Sea Surface Temperatures and Glacier Ice

Oceans cover some 70 percent of the earth's surface, but only a tiny handful of our long-term climate records truly reflect oceanic temperatures. In the North American sector alone, the number of continental land-based stations exceeds the number of oceanic stations by at least 3,000 to 1.

Yet it is fairly clear that regional climates are often controlled by the temperatures of the oceans. In the midlatitudes, land areas (such as the West Coast of North America) that face cold oceans tend to have cool, wet climates in the immediate vicinity of the water, but inland, to the lee of the coastal mountain ranges, dry, even desert, conditions prevail. Similarly, land masses (such as the

73

eastern part of North America south of Newfoundland or mainland China) that face warm oceans tend to experience very humid climates with long summers.

It is also clear that major changes in the temperature of the oceans will have a profound effect on the climate of most continents, and yet the large liquid portion of the globe is poorly monitored by stationary thermometers. Further, almost all measurement sites are on islands. The notion of continental plates and sea floor expansion dictates that almost all islands in the open ocean are of volcanic origin and that they have sharp elevation gradients that may not reflect oceanic climate very well at all.

Santa Helena—noted earlier as an important ocean station that is obviously in error—fits into that category, and, because of its remote location, its temperature is used to indicate that of much of the mid–South Atlantic Ocean. Other single stations that have inordinate influence on the global temperature history include Hawaii (in the Central Pacific) and Tristan de Cuhna (in the far South Atlantic). In fact, the 70 percent of the planet that is covered by water is monitored by only a handful of island stations.

It is apparent that the placement of those few stations, which are so important to the global record, significantly determines the temperature recorded for large areas. One alternative is to use temperatures recorded from traveling ships, but, as noted earlier, those readings have to be artificially corrected for measurement biases.

Probably the most carefully collated of the shipboard-measured sea surface temperatures (SSTs) was recently published by J. W. Bottomley of the U.K. Meteorological Office, Reginald Newell of Massachusetts Institute of Technology, and others. The SST records (see Figure 5.13) correspond remarkably well to land readings after 1900 with respect to the timing of warming and cooling. The magnitude of change in SSTs, however, is about half that of land temperature changes. What is most notable is that this record does not show any very long "lag" behind the land record, so that if the oceans are holding back the warming, they are definitely not doing so at the surface layers. The SST readings also do not show the very warm values in the Northern Hemisphere at the end of the record (after 1980) that appear in the primarily land-based records.

There is an additional corroboration of the SST data from the unlikely location of the Himalayan ice field. Lonnie Thompson and

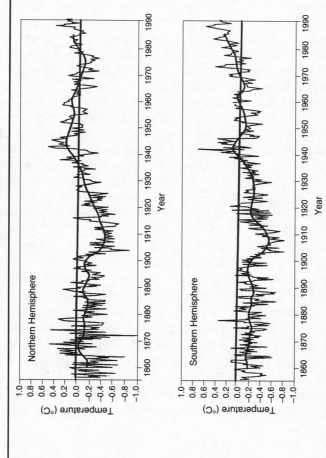

Figure 5.13
SEA SURFACE TEMPERATURES

NOTE: The SST record of Bottomley and others (1990) behaves much like the land record between 1900 and 1980, except that the magnitude of its changes is about half of what it is in land-based histories.

his many colleagues at Ohio State University have been measuring the concentration of two isotopes of oxygen trapped at various levels in glaciers around the world. The ratio between the two isotopes provides a fairly good estimate of the temperature of the water *source* that eventually fell as snow and fed the glacier. What Thompson found was very similar to the Northern Hemisphere's history, in terms of both the land records through 1980 and the entire SST history: a rapid warming in the early 20th century, followed by a period of cooling that ran roughly from the mid-1930s into the 1970s, followed by another warming. As shown in Figure 5.14, the warmth at the end of the record appears equal to the maximum observed in the 1930s.

In a repetition of a now-familiar news story, the press reported that Thompson's *Science* article showed evidence of global warming caused by an enhanced greenhouse. One scientist who was interviewed attested to that fact by saying that continental locations (such as the ice field) should warm up faster than the oceans, and the warming in Thompson's record attested to that. Of course, the oxygen ratio–derived temperatures reflect the temperature of the water source for precipitation (in this case, the Indian Ocean), not that of the land.

Parenthetically, Thompson testified before Sen. Albert Gore, Jr. (D-Tenn.), early in 1992 that tropical glaciers in South America were receding rapidly, a story that also received considerable press play. Because of their tropical latitude, those glaciers are confined to the highest elevations, some three miles above sea level. The news stories chose not to mention the lack of much warming in the temperatures of the oceans that ring the tropics, or that the glaciers were receding (at a slower rate) even when the planet was *cooling* from 1940 through 1975. Further, Thompson emphasized that glaciers in the high-latitude, polar region—where warming, according to the GCMs, should be most profound—show little if any consistent melting.

Another glacial record (Figure 5.15), published in 1988 by Fred Wood in *Arctic and Alpine Research*, tells a story that seems even more opposed to the Popular Vision of what is happening to the atmosphere.

It is well known that mountain glaciers are very sensitive to slight changes in temperature. For example, paintings of the Alps in the

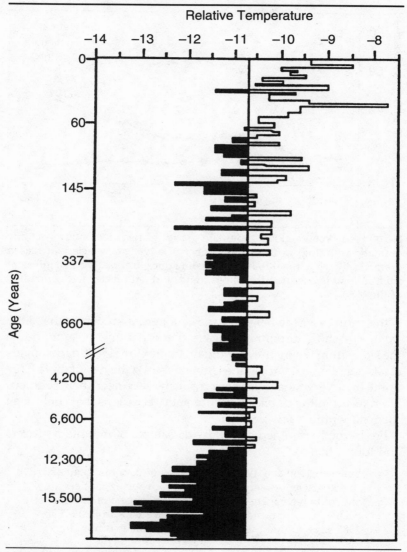

Figure 5.14
THOMPSON'S HIMALAYAN OXYGEN RECORD

NOTE: Lonnie Thompson's temperature history from the oxygen ratios in the Himalayan ice cap appears similar to the SST records.

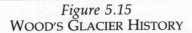

Figure 5.15
WOOD'S GLACIER HISTORY

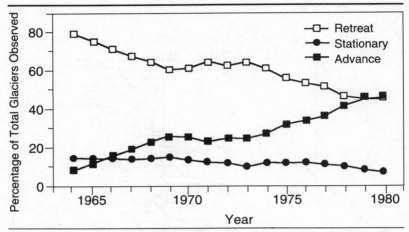

NOTE: Fred Wood's global mountain glacier history does not reflect any dramatic warming; in fact, the number of advancing glaciers increased dramatically while the number of receding ones dropped by an equivalent value. Wood has written that those trends did not reverse in the first half of the 1980s.

17th century—at the height of the cool period known as the "little ice age"—show termini of glaciers that are a mile or more down the slope from where they are today. In the early 1960s park rangers at Glacier National Park in the United States, which contains a large number of very small glaciers obviously very near their southern limit, lectured the public about dramatic recessions that could melt them all within a few years.

Fred Wood examined a worldwide sample of mountain glaciers and found that

> between 1960 and 1980, on the basis of data for about 400 to 450 glaciers observed each year, advancing glaciers are shown to have increased from about 6 percent of observed glaciers to 55 percent. . . . Preliminary data from 1981 to 1985 suggest that the mixed glacial regime is continuing.

Because so many features of glaciers are very sensitive to slight temperature changes, it appears that there is no dramatic greenhouse warming that has somehow been missed by our large area temperature records. In addition, glaciers may yet serve as the arbiters of the difference between the ground-based temperature records of the 1980s, which in the Southern Hemisphere show dramatic warming relative to the satellite data that began in 1979.

Temperature Histories: Overall Conclusions

We are left with the following conclusions about global and hemispheric temperatures. The general characteristics of the various temperature records compiled between 1900 and 1980 appear to be consistent: A fairly rapid warming, most pronounced in the Northern Hemisphere, of approximately 0.4°C (0.7°F) occurred between approximately 1920 and 1940. That warming was followed by a period, which lasted through the mid-1970s, during which the temperature cooled by 0.2°C (0.4°F) in the Northern Hemisphere only. After 1980 there is a remarkable discrepancy between the ground-based and the satellite data for the Southern Hemisphere, with the ground record showing warm temperatures that are virtually nonexistent in the record sensed from space. That difference remains unexplained.

The overall global warming of the past 100 years ranges between 0.3°C and 0.5°C (0.5°F and 0.9°F), depending on what adjustment is made for local urban warming. At least half of that warming occurred before two-thirds of the enhancement of the major greenhouse gases took place. The Southern Hemisphere, which should warm least and slowest, in fact shows a more "greenhouse-like" pattern than the Northern Hemisphere, which displays no warming within the normal bounds of statistical significance in the past 55 years.

The temperature history of the planet supports neither the Popular Vision of climate apocalypse nor the mid-1980s GCM forecasts of global warming, either in pattern or in magnitude. Yet both the Vision and those forecasts are driving the most comprehensive experiment in the central planning of energy in the history of the human race.

6. Ghosts in the Apocalypse Machine

"The Data Don't Matter"

It is now generally acknowledged that the mean history of global, hemispheric, and regional temperature simply is not consistent with either the Popular Vision or the mid-1980s GCMs. The reasons are manifold but generally include the lack of strong warming in the Northern Hemisphere and the apparent inversion of the hemispheres, with the Southern Hemisphere behaving in a more "greenhouse-like" fashion than the northern, even though the Southern Hemisphere should be less greenhouse-like. Also important is the pronounced lack of high-latitude winter warming that should already have occurred.

Suppose, though, the Northern Hemisphere had a "natural" and "random" cooling that occurred just as the increase in the trace gases became important (after 1950), and the expected greenhouse warming was canceled by that fluctuation. Although that supposition sounds preposterous, it is precisely the argument that has been tendered to explain the lack of warming. No causation is implied or suggested, and the argument resorts to the complete unknown (a random perturbation) in an attempt to explain what is known (a lack of warming).

The genesis of that argument is interesting and dates back to pioneering work by Edward Lorenz of the Massachusetts Institute of Technology. Lorenz was one of the first to attempt to model climate, and his work began before the advent of supercomputers. Thus, he was initially constrained to relatively simple designs that could be accommodated on the machines of the time. That was especially true because he wanted to see if he could simulate the behavior of the climate for centuries, a task that could block up a computer for days. For what it is worth, we have yet to successfully reproduce the behavior of the past 100 years from first physical principles.

Climate models—from Lorenz's early simulations to the most complicated GFDL ocean-atmosphere version—are nothing but a series of interacting differential equations whose parameters are determined from known physical quantities or, in some cases, guessed. Sometimes the various output factors, such as global mean temperature, sum up to produce a warm period; sometimes they cancel and produce the average temperature; and sometimes they add negatively and simulate a cold period.

NASA's Jim Hansen ran the Goddard Institute for Space Studies (GISS) climate model with no greenhouse enhancement. In his paper published in 1988 he found that a "random" cooling of approximately 0.4°C (0.7°F) occurred during years 50–75 of a 100-year run. The purpose of the experiment was to estimate how much climate can vary without any changes in the overall driving variables such as solar intensity, the greenhouse effect, or planetary reflectivity.

Is the GISS model up to this task? According to a 1991 paper by Stanley Grotch of the Lawrence Livermore Laboratory that appeared in a recent review volume on climate change edited by GCM modeler Michael Schlesinger of the University of Illinois, the current error in its estimation of Southern Hemisphere temperatures poleward of 70° is approximately 12°C (22°F), and the sea surface temperatures are artificially set "to ensure a match of sea surface temperatures to their [known] climatological averages." Thus, the temperature of 70 percent of the earth is fed in as the "right" answer, and errors in excess of 10°C (18°F) are still generated over the remaining 30 percent.

So, if we make the wildly optimistic assumption that the model truly mimics reality, such a "random" cooling, if it occurred during the past 40 years, would be sufficient to cancel out much of the warming that most people feel should have taken place in the Northern Hemisphere.

That argument is at the core of the frequently made assertion that "the [observed] data don't matter" as far as greenhouse policy is concerned. Chris Folland of the United Kingdom Meteorological Office, which is chaired by Sir John Houghton (senior editor of the IPCC report), uttered those very words during a presentation on observed climatic change at a meeting of climatologists in Asheville, North Carolina, on August 13, 1991. The slide that was on the screen at the time was Figure 5.7, showing that around 1950 the Northern Hemisphere stopped warming at the rate established *before* the greenhouse enhancement was consequential. The presenter was incredulous and asked Folland to repeat his statement so that the entire audience could hear, and Folland again said, "The data don't matter."

When pressed, Folland replied that random runs of non-greenhouse-enhanced climate models (Figure 6.1) produce random coolings of 0.4°C or 0.7°F (as in Hansen's experiment) and that such behavior could explain the lack of warming of the Northern Hemisphere as the greenhouse gas levels went ballistic. "Besides," he added, "we're not basing our recommendations [for immediate reductions in CO_2 emissions] upon the data; we're basing them upon the climate models."

Again, if we assume that the GISS model is reliable, what is the likelihood that Folland's assertion is correct? The 100-year "background" (unperturbed) run shows one period of approximately 25 years in which cooling takes place. That is one-fourth of the time. But an equal and opposite random warming is just as likely; otherwise, the model would continue to create unidirectional climate

Figure 6.1
HANSEN'S RANDOM RUN

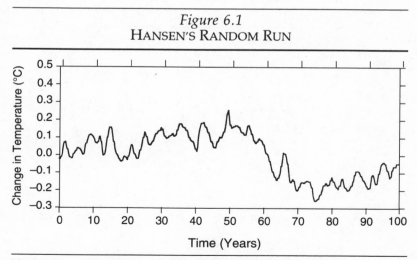

NOTE: Mean global temperatures were simulated by NASA's model run without any greenhouse enhancement. Note the sudden drop of 0.4°C (0.7°F) for more than 20 years.

changes that we know are not true. The likelihood that any quarter century would show such a cooling is, therefore, only 1 in 8. The Northern Hemisphere shows no statistically significant warming in the past half century (on a linear trend basis). Therefore, the odds of such a suppression in the face of the trace gas increase are even longer than 1 in 8—a *very* conservative estimate would be 1 in 10.

Thus, even assuming a model is correct when it has known gross errors, the chance that a random ghost in the climate machine is suppressing the warming of the Northern Hemisphere is only 1 in 10. This is then cited as the basis for what will be the greatest experiment in central energy planning in history.

Other GCMs have also been run randomly. Figure 6.2 is an illustration from a 1991 *Journal of Climate* paper by David Houghton, Bob Gallimore, and others. It shows only random fluctuations of one- or two-tenths of a degree. Who can judge which ghosts are more real?

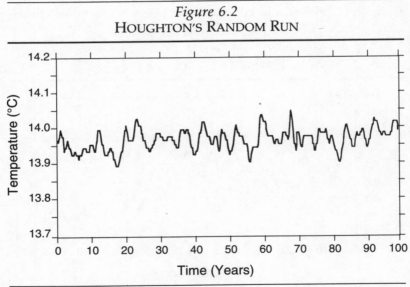

Figure 6.2
HOUGHTON'S RANDOM RUN

NOTE: The random run is of a GCM by Houghton and Gallimore. Those fluctuations are only half as large as the ones in Hansen's model. Which, if either, of the two is correct?

Monster in the Depths?

Now that we have established that the likelihood of a "random" cooling sufficient to have hidden the expected warming of the Northern Hemisphere is very small, what other arguments have been generated to bolster the Popular Vision?

The other ghost in the machine is the monster from the depths. Its form and substance are about as definable as the random cooling hypothesis, and there is only one piece of evidence—wholly uncomprehended and unexplained—supporting its existence. In a truly bizarre twist, if that evidence is indeed the ghost, the world is destined for a dramatic warming that it simply cannot stop with any economic intervention.

What has been established is that global sea surface temperatures are not warming as much as they should unless something is either preventing warming or holding it back. If the former is true, it would be instructive to know what the agent is; if the latter is true, the agent must be in the deep ocean, because shallow SSTs show very little lag behind land temperatures. Unfortunately, the "climatology" of the deep ocean is so poorly understood that we can barely guess at the form of some of the important equations.

Figure 6.3 shows the change in temperature between the late 1950s and the early 1980s for various depths in the North Atlantic and Pacific Oceans. The data were taken for submarine warfare purposes by the former Soviet Union, because several aspects of submarine operations are highly temperature dependent.

The upper layer of both ocean regions—roughly the upper 2,000 feet—shows a cooling of as much as 0.2°C (0.4°F). That is consistent with the 3°C (5.4°F) winter cooling of the Arctic surface temperatures in the North Atlantic that Rogers noted in 1989. It is also consistent with some of the regional changes in SST noted in Bottomley's (1990) sea surface temperature atlas that was cited in the last chapter. Beneath 2,000 feet, the North Pacific Ocean shows no change in temperature between 1950 and 1980.

The story is much different for the lower levels of the North Atlantic. From 2,000 feet down to 10,500, temperatures have warmed. The magnitude of the warming is not great—less than 0.2°C (0.36°F)—but the heat capacity of water is tremendous compared to that of air. Further, the depth of the warmed layer is so great that if its warming were distributed to the atmosphere, surface temperatures would warm up over 4°C (7.2°F).

Figure 6.3
DEEP OCEAN TEMPERATURES

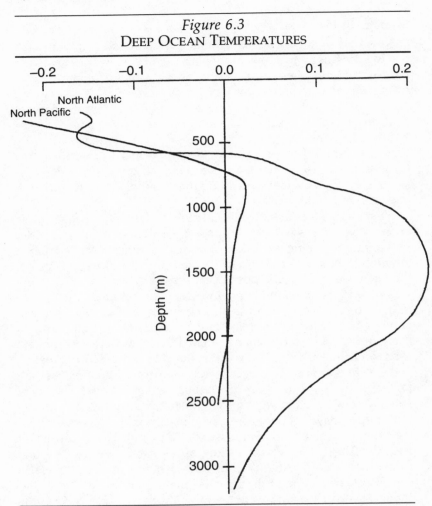

NOTE: The differences in ocean temperatures at various depths in the North Atlantic were measured between the 1950s and the 1980s.

The usual scientific questions are, first: "What is causing the warming?" Answer: no one knows, because the warming shows up on only one of the two northern oceans. Next: "Are the data reliable?" (We are dealing here with very small temperature changes at very high pressures.) Answer: no one knows. Third: "Is such behavior typical of those depths?" Answer: no one knows.

Fourth: "If the warming is caused by an enhanced greenhouse effect, why is it buried?" Answer: no one knows, but an answer in the affirmative along with a plausible mechanism would probably revolutionize our understanding of the problem.

Again we resort to the unknown to explain the known lack of warming. That point of view, which virtually admits that "something" has to be amiss with the current forecast, was recently summarized in the National Academy of Sciences' 1991 report, *Policy Implications of Global Warming*. The fifth of the five "scientific conclusions" in the report follows.

> Several troublesome, possibly dramatic, repercussions of continued increases in global temperature have been suggested. No credible claim can be made that any of these events is imminent, but none of them are precluded.

The report also discusses recommended policies that have become the subject of considerable contention. However, one has to ask, "Given the lack of expected warming, and the appeal to the unknown, are those policies based on ghosts?"

7. Climate Change: The New Vision

While the preceding chapters make it clear that "the details have yet to be worked out," there are a few generalizations that we can make with some confidence about what happens when atmospheric CO_2 increases. One is that, although warming may be hard to detect at some surface locations, the upper regions should cool. The GCMs say that cooling should be confined to altitudes skyward of 40,000 feet, but Jim Angell of the National Oceanic and Atmospheric Administration (NOAA) has demonstrated that the lower limit to cooling is around 25,000 feet, which is well within the lower atmosphere's active "weather zone." Even though surface warming may be mitigated, at least in part, by confounding factors such as an increase in cloudiness, it is difficult to envision a mechanism that would preclude a cooling of the higher altitudes.

100 Percent Chance of More Rain

If the temperature of the atmosphere were everywhere constant with height, the planet would be a lifeless desert, because there would be no rain. Fortunately, that is not the case (and cannot be) thanks to some rudimentary principles of gaseous behavior that have been known for centuries. Their implication—that the more the temperature of the weather zone cools with height, the more it rains—has interesting ramifications in a world with an enhanced greenhouse effect.

The region from 25,000 to 60,000 feet represents the upper half of the "active" region of the atmosphere, where up-and-down motions induced by temperature changes give rise to weather systems great (the thousand-mile cyclones that characterize our winter) and small (the individual summer thunderstorms on which midlatitude agriculture is so dependent). The more substantial the temperature contrast is between that level and those underneath, the more energetic are the attendant weather systems. If the contrast is great in the North American or Eurasian winter, the jet stream is stronger and the cyclones are more vigorous. If the same is true in the late spring or summer, the number of thunderstorms is likely to rise. Those observations seem somewhat at variance with section VIII.B of the outline of the Policymakers Summary: "Our forecast models give no clear indication of whether or not hurricanes or midlatitude (winter) storms will increase or decrease in frequency or intensity."

That scenario plays out a bit differently in the tropics. Near the earth's "thermal equator" an increase in surface warming or upper cooling will, in fact, enhance the already frequent thunderstorms that characterize the area. But what goes up (the rising air that forms thunderstorms) must come down, and the downward current is as dry as the ascending current was wet.

That combination of large-scale upward (cooling and precipitating) motion and downward (warming and drying) motion is responsible both for the tropical rain forests (which tend to be near the thermal equator) and for the surrounding deserts. The more the thunderstorm mechanism is enhanced (by surface warming and/or upper cooling), the more the deserts will also be reinforced.

Another way to make it rain more is simply to put more moisture in the atmosphere, even if the upper temperatures remain unchanged. There is only one way to do that in meaningful quantity: warm up the surface temperature of the major source of water, the oceans.

According to long-term records, global precipitation is increasing. Further, in the circumscribed areas—many of the world's deserts—where we would expect some decreases, there also appears to have been a decline. That change is very evident in precipitation records published by R. S. Bradley and his coworkers in the *Science* in 1987 (see Figure 7.1). The U.S. record (Figure 7.2) also reveals (contrary to the blizzard of news reports about this drought or that crop failure or the pictures of dead chickens that accompany every summer heat wave) that the 1980s, followed by the 1970s, were the wettest decade in our reliable weather history. The preponderance of recent years with above-normal precipitation can be appreciated by covering up all of the pre-1970 values and noting how many of the succeeding ones are above the century average.

Those considerations present an interesting problem: if rainfall is increasing, which of the three mechanisms that can cause it (sea surface temperature warming, land surface warming, or upper atmospheric cooling) is responsible?

Figure 7.1
BRADLEY'S PRECIPITATION RECORD

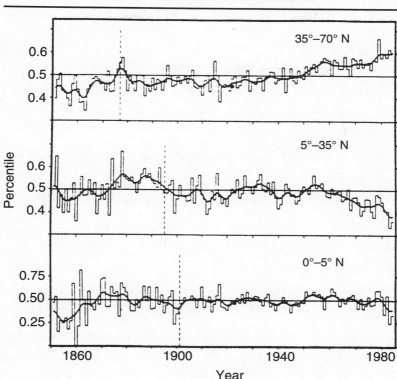

NOTE: Bradley's (1987) regional analyses of rainfall show the expected increase in the midlatitudes and decrease in the already dry desert regions.

It is well established that SSTs have not risen very much at all, as shown in the previous chapter. Any increase is surely too little to create any statistically significant change in global precipitation. Further, the Northern Hemisphere's surface temperature record, which is primarily from land-based stations, shows no statistically significant change in temperature in the past half century. That leaves upper cooling as the remaining candidate.

In fact, significant drops in stratospheric temperatures have been measured over the past 10 years in association with the Antarctic late winter–early spring ozone depletion (the erroneously named

Figure 7.2
REGIONAL PRECIPITATION RECORDS

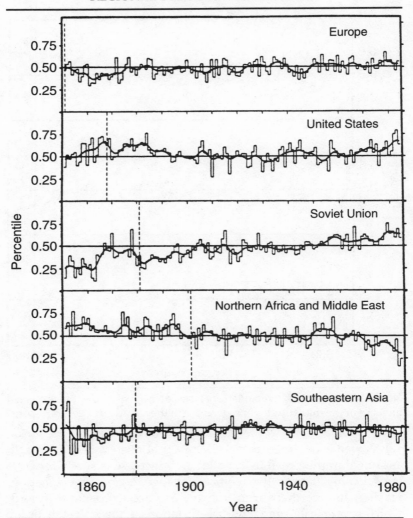

NOTE: Contrary to the plethora of scare stories, there hardly has been a decrease in U.S. precipitation. In fact, the 1980s were the wettest decade in our record, which is considered reliable after 1870 (dashed line).

"ozone hole"). Some researchers are now speculating that a portion of that depletion is being driven by greenhouse enhancement. But the stratosphere is above the earth's "weather zone" where precipitation would be enhanced by an upper cooling. Nonetheless, in work that others and I have published (Michaels et al. 1990), there is evidence that temperatures upward from approximately 18,000 feet, which can enhance precipitation, also have dropped in the past 40 years.

Thus, we are forced to ask the following question: "If there is some evidence of an upper cooling since 1950, why do we not see much surface warming, a lack that is painfully obvious in the SST and Northern Hemisphere records?" About the only plausible answer is that there has been some increase in *cloudiness*, which could have been caused by upper cooling, by an increase in the number of cloud-forming particles in the lower layers, or perhaps by just a slight increase in lower cloudiness due to increased rainfall. The last phenomenon can be appreciated by noting the low-level stratus and fog that tend to occur after a storm. Whatever its cause, the cloudiness does not seem related to a profound surface warming. And is there any evidence that clouds are increasing?

Increasing Cloudiness: Ecological Implications

The prime moving force behind the Popular Vision of apocalypse is *not* a rise in temperature but an increase in the rate and amount of evaporation from land surfaces. In fact, by itself, a rise in temperature might actually stimulate plant growth (it would certainly lengthen the growing season). Some of the world's most productive ecosystems are found in regions of very high mean temperature.

Some of the least productive ecosystems—the low-latitude deserts—also experience very high mean temperatures, and the difference between the deserts and lush forests is due to rainfall and evaporation. Overall mean temperatures are not very different.

We have already seen that one of the hallmarks of an enhanced greenhouse is an increase in global precipitation, especially near the earth's thermal equator and in the agriculturally productive midlatitudes. The logical question is, then, "What drives the calculations of agricultural disarray and dying trees?" Those are forecast to occur because the evaporation rate is projected to increase to a level that more than compensates for the rise in rainfall.

Over the world's productive agricultural regions, evaporation is primarily restricted to daytime hours when the sun beats down. At night, the opposite—condensation—tends to dominate, as evinced by the morning ritual of cleaning water (frozen or otherwise) off the car's windshield before going to work. If cloudiness increases, then daytime evaporation drops even as rainfall goes up. An additional benefit is that clouds, because of their own greenhouse-enhancing effect (water vapor, not CO_2, is by far the most abundant greenhouse gas in the atmosphere), tend to reduce the nighttime cooling rate, which will lengthen the growing season.

The profound effect of clouds on night temperatures is easily appreciated by readers who live in regions frequently subject to cold airmasses of polar origin (that is everywhere east of the Rocky Mountains in North America). When a (usually misnamed) Siberian Express first appears, initial conditions are clear and windy, but as the center of the airmass approaches, a period of calm develops and lasts for 24 hours or so. Often, an increase in high- and mid-level cloudiness occurs near the end of the calm even though surface winds have yet to switch around to a more warming southerly direction. If the clouds come in at night, the temperature immediately stops its normally rapid nighttime fall—even though no warmer air has been pumped into the region. The sudden increase in heat retention by clouds has dramatically held up the night temperature.

In fact, the general climate characteristics accompanying an increase in cloudiness should be increased rainfall, decreased evaporation rates, warmer nights, and longer growing seasons—hardly the Popular Vision's prescription for climate apocalypse and certainly not resonant with the overall tenor of the Policymakers Summary on climate change.

But what of GCM predictions of increased frequency of drought in major agricultural regions of North America? Those predictions form a considerable part of the basis for concern about climate change. Suffice it to say that, to save on computing time—an unfortunate necessity with huge models and today's cybernetic technology—most of the mid-1980s models did not have explicit 24-hour day and night (evaporation and condensation) cycles. Hansen's GISS model was an exception, but to date it has misestimated the ratio of night-to-day warming by 9 to 1 in the Northern Hemisphere.

As a result, the models cannot successfully calculate the implications of an increase in cloudiness.

Clouds and Climatic Change

The temperature of the earth's surface is determined by the amount of solar radiation coming in minus the amount that is reflected, and that energy balance is adjusted by the greenhouse gases.

As it stands today, the earth reflects a bit more than one-quarter of the incoming solar radiation, and clouds are one of the prime reflectors. In fact, a mere 2 percent change in global reflectivity (the numbers vary a bit, depending on the assumptions that are made) would create enough cooling to totally offset the warming associated with an effective doubling of CO_2. While the numbers again vary some with the assumptions, that amounts to roughly a 4 percent increase in average global cloudiness.

Increasing Cloudiness: The Data

Because a small increase in cloudiness could spell the death of the Popular Vision of imminent climate catastrophe, it is worthwhile to examine the scientific literature on the subject.

In a 1987 paper in the *Journal of Climate and Applied Meteorology*, William Seaver and James Lee reported a statistically significant decrease in the number of cloudless days over the entire United States. The periods of comparison were 1900 to 1936 and 1950 to 1982.

However, early observations were not made by individuals specifically trained for those duties. Airport personnel, however, are a different lot because they are charged with safety maintenance that is highly weather dependent. In his 1990 article in the *Journal of Climate*, James Angell examined records taken by those observers (Figure 7.3) during the latter period (1950–82). He reported that in

the United States cloudiness increased by 2.0–1.3 percent between 1950–68 and 1970–88 (corresponding to a percentage increase of 3.5 percent since the average cloudiness was 58 percent . . . during 1950–88). The increase in cloudiness was close to 2 percent in all regions of the country and was (statistically) significant at the 5 percent level in all regions except the southeast.

Figure 7.3
ANGELL'S CLOUDINESS RECORD

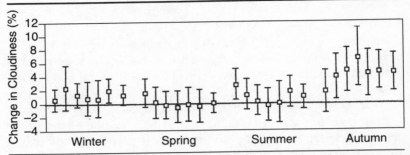

NOTE: Jim Angell (1990) of NOAA recorded seasonal cloudiness changes over various regions of the United States. His figures are the difference between the 1950–68 average and that for 1970–88. Angell reported that the annual differences were statistically significant in all locations except the Southeast. The *lack* of a spring increase turns out to be very important, as noted later.

The weather services of some countries deploy instrumentation that measures the amount of time that the sun is not obscured by clouds, and German climatologist Gerd Weber examined those records for the German Republic.

> Sunshine generally decreased (since 1950). . . . For some stations, the loss amounts to more than 300 hours per year, more than 18 percent of total received sunshine. Mountain tops . . . show the greatest decline; this may result from increased cyclonic episodes over Central Europe. The decrease . . . was not accompanied by increased precipitation or by an increase in the number of days of precipitation.

The fact that increases seem to be magnified near mountaintops suggests a change in the low-level stratocumulus cloud type, which is one of the varieties most effective at global cooling. Because there are no concomitant increases in precipitation in the region, the amount of moisture in the air has probably not increased. It is also likely that the winds have not changed in a fashion that would bring more moisture ashore, that the ocean surface has not warmed, and, therefore, that evaporation has not increased.

All of this discussion leads to the proposition that there has been an increase in some chemical or physical species in the air and that the change is capable of increasing cloud frequency without necessarily requiring more moisture for cloud formation.

Australian climatologist Ann Henderson-Sellers examined records—primarily military and governmental—for all of North America and found dramatic cloud increases, averaging around 8 percent this century (see Figure 7.4). Even before she wrote that paper, she commented in an earlier article in *Climatic Change* that historical studies of cloud frequencies had shown that they had increased during the modest warming of the past 100 years and that this observation was already in conflict with what computer models of climate predict.

Perhaps the most comprehensive examination of cloudiness was recently published by Steven Warren and his coworkers (Warren et al. 1988) at the University of Washington (see Figure 7.5). They examined the literally millions of records of cloud cover taken every three hours by shipboard observers.

There are some problems with the data—including a measurement rubric that changed in the early 1950s from "eighths" to "tenths" in the estimate of total sky coverage—but the overall trends are unmistakable. Even if we disregard the data before 1950, there is an overall increase in Northern Hemisphere cloudiness of 2 percent from then through the early 1980s, which is the current termination point of the analyzed record. There is also an increase in cloudiness in the Southern Hemisphere, although the magnitude of that increase is about half of that for the Northern Hemisphere.

Warren's (1988) record has also been divided into various cloud types, latitude bands, and seasons. What appears remarkable are the large increases observed in the industrial latitudes of the Northern Hemisphere, along with the tendency for large increases to be in the low-level stratocumulus. Not only is that cloud type very effective at cooling, it is the type that might be enhanced by industrial activity.

Increasing Cloudiness: A Hypothesis

We cannot deny that greenhouse gases have increased dramatically on a global scale ("all scientists agree"), and yet there seems to be scant evidence of any warming that is consistent with the Popular Vision. Further, the disparity between the two hemispheres' temperature histories begs for at least a partial explanation.

Figure 7.4
HENDERSON-SELLERS'S RECORD OF CLOUDINESS

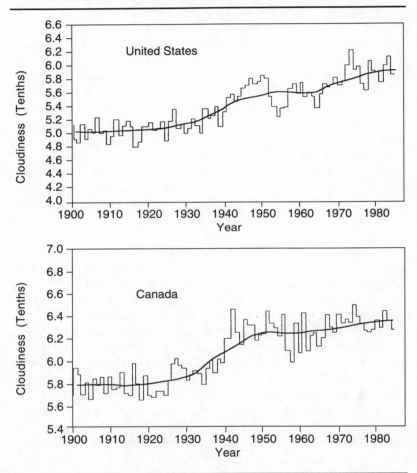

NOTE: Ann Henderson-Sellers's records of changes in cloudiness (in tenths) for the conterminous United States (top) and Canada (bottom) were published in *Global and Planetary Change* (1990). They were also included as a "supporting contribution" to the IPCC report. No mention of the implications of her data was made in the Policymakers Summary.

Figure 7.5
WARREN'S RECORD OF CLOUDINESS

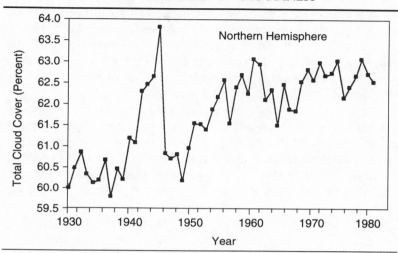

NOTE: These cloudiness changes since 1950 in the Northern Hemisphere were measured aboard ocean-going vessels. Data were published by Steven Warren and his coworkers (Warren et al. 1988) at the University of Washington.

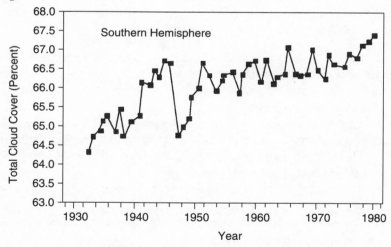

NOTE: Although the data are more sparse and less reliable, the Southern Hemisphere cloud increase since 1950 is about half of the magnitude of that observed in the Northern Hemisphere.

Hypothesis: A confounding factor is being introduced into the atmosphere and is acting primarily in the Northern Hemisphere. If that is true, the factor cannot be a long-lived species, because relatively little lower atmospheric air exchanges between the two hemispheres each year. If the residence time of the confounding factor were, say, as long as carbon dioxide's (50 years or longer), it would exert an equal influence on both hemispheres.

The most likely candidate is the family of anthropogenerated particulates: dust and finely divided aerosols that are the by-products of civilization. Coal and petroleum are especially rich sources, because of their minute concentrations of sulfur, which becomes sulfur dioxide in the combustion process. Many of those emissions are ultimately oxidized to form sulfate particulates. Dust from agricultural activity may also be important.

The aerosols are effective cloud-seeding agents, also known as cloud condensation nuclei (CCN), that brighten clouds by increasing the number of small droplets or create clouds where there were none. The net result is increased reflection of solar radiation, which reduces daytime temperatures. At night, the heat-trapping effect of enhanced clouds tends to raise the temperature.

If clouds were increasing because of anthropogenerated particulates, we would expect the following:

1. Nights should warm from both the increase in greenhouse gases and the increase in cloudiness;

2. There should be a counteraction of daytime warming because of the increase in clouds;

3. There should be a consequent decrease in the daily temperature range, or in the difference between high and low temperatures;

4. The warming (nighttime effect) from clouds should be pronounced on (long) winter nights;

5. The cooling (daytime effect) from clouds should be pronounced on (long) summer days;

6. The warming (nighttime effect) from clouds should be attenuated on (short) summer nights;

7. The cooling (daytime effect) from clouds should be attenuated on (short) winter days;

8. The effects should be concentrated in the industrial Northern Hemisphere; and

9. Cloudiness should be enhanced near the particulate source regions of North America and Eurasia.

How many of those implications of enhanced cloudiness are verified by the data? Can the observed changes be explained solely by anthropogenerated particulates, or are those changes also a natural consequence of a greenhouse enhancement? What constitutes a sufficient number and pattern from which to conclude that warming has been mitigated or channeled into a pattern that radically alters the Popular Vision?

It's Not New

The idea that anthropogenerated aerosols are countering an enhanced greenhouse is hardly new. Reid Bryson, now a very active emeritus professor at the University of Wisconsin (Madison), holds the patent on the idea, but he was a victim of very unlucky timing. Two decades after he founded the Meteorology Department at Wisconsin, Bryson started the Institute for Environmental Studies, which was the first of the interdisciplinary groups specifically designed to wrestle with the problems of the whole earth and with the interactions between the dynamic atmosphere and ecosystems.

One of his meteorology classes—Past Climates and Climatic Change—became the launching pad for several prominent academic careers. Even as early as the 1960s, Bryson would go through the standard derivations of the energy balance equations, the forcing of the jet stream, and planetary mean temperatures, including the effects of a greenhouse enhancement. Then—almost 30 years ago—he would turn to the class, chuckling, all eyebrows and Coke bottle glasses, and say, "We have a problem here. Why isn't it warmer?"

His answer was the "human volcano"—particulates produced by human economic activity—a concept very close to what we now hypothesize is mitigating the warming. In fact, by the mid-1970s Bryson had parameterized his theory so well that it explained a massive amount of the observed temperature history of the planet, but publishing the paper in a first-line journal took years. And when it did appear in 1976, the GCM-driven warming craze and the explosion of hot dollars had just started.

The Difference between Night and Day

The ecological implications of a greenhouse warming modulated by an increase in low-level cloud are profound, and they dramatically alter the Popular Vision of climate apocalypse. The apocalyptic vision of withered crops, burned forests compromised by poor growth because the climate has changed so fast that they have been unable to follow, and increased sea levels inundating the world's major ports and cultural centers is driven by daytime warming and the melting of vast areas of land ice.

Low-level clouds are so bright that they provide a net cooling, even as they warm the nights. Thus, daytime temperatures would cool if only the clouds dominated, while the effect might be neutral during the day if the greenhouse enhancement were sufficient to counter the cloud effect.

In fact, within some broad limits, land warming is ecologically inconsequential unless it induces an increased amount of moisture stress on the world's biota. *Evaporation*, rather than temperature, drives down models of forest and plant growth when subjected to greenhouse scenarios. What the Popular Vision models sense is that, even with a projected increase in global rainfall, the rise in temperatures will increase evaporation so much that plants will have less available moisture than they did before the world warmed up.

Many of our agricultural plants frequently undergo severe moisture stress when evaporation rates skyrocket. Even in the heart of the corn belt, with its usually high humidity, it is not uncommon on sunny, hot days for leaves to shrivel as the water concentration in the tissue drops below some critical value. When that occurs, the stomata—pores through which moisture and CO_2 exchange—close, shutting the doors to the nutritionally critical CO_2 while preventing much further loss of water. As a result, if plants are under frequent moisture stress, they cannot take in enough CO_2 and fix it in the form of plant carbohydrates. The normal processes of life maintenance, including respiration and consumption of nutrients, then become so demanding that net plant growth becomes neutral or even negative.

If warming is confined primarily to the night, the relative increase in evaporation is lessened, and the main cause of plant moisture stress is diminished. There is an additional benefit: the growing season (the period between the last spring frost or freeze and the first in the fall) lengthens. Almost all very early (fall) or late (spring) frosts occur near dawn after clear, calm nights dominated (in North America) by airmasses of Canadian or Arctic origin—the coldest of which, according to Larry Kalkstein (Kalkstein et al. 1990), have warmed.

Critics of that view have noted that night warming—because of the associated propensity for longer freeze-free periods and less severe winters—also means either that insect populations will reach higher numbers or that infestations will be more persistent. While that is probably true, a general rule is that the lower the difference between the daily high and low temperature, the more lush the standing vegetation.

Although there are obvious local exceptions, the beneficence of that phenomenon is quite apparent from a global perspective. Polar regions—where at the extreme there is only one very cold night a year—are devoid of vegetation. Subtropical latitudes, from approximately 20 to 30 degrees on each side of the equator, also have very large daily temperature ranges and are the home of the world's great deserts. On the other hand, the most lush environments, the rain forests that dominate the western coast of North America, along with the great tropical forests, are characterized by the lowest daily ranges in temperature, in part because they are very wet.

The apocalyptic prediction of rises in sea level caused by melting of land ice meets a similar end when subjected to night, rather than day, warming. The high-latitude regions, with the great continental ice sheets, tend to be very cold—around −40°—during the long polar night. Thus, any warming foreseen by even the wildest projections does not bring the temperature of those regions close to freezing. If cloudy days prevent much summer warming (and even the GCMs do not project a lot of high-latitude warming then), it will become very difficult to melt much ice and send the seas coursing through Manhattan or lapping at the Washington Monument to fulfill the Popular Vision.

Melting of floating sea ice, such as the North Polar ice cap, is a different story. Its melting would not change the water level any more than a cube's melting changes the level of liquid in a glass.

Nonetheless, sea level could rise some if only the nights warmed, because water expands slightly as it warms up. But that residual rise would be only about one-half of what would occur if the warming were equally distributed between day and night and if it melted land ice. Why do I use the word "could," given the known thermal expansion of water? Because it is unclear how much water will be locked up by the spreading of Antarctic (and possibly Greenland) ice as the atmosphere warms enough to hold more moisture and that moisture is precipitated as snow. Research cited in the previous chapter suggests that the expansion of ice, at least for the first few degrees of high-latitude warming, will be significant.

Finally, a night warming ipso facto is primarily a winter warming, because nights are longer in the winter. Thus, we might be facing the prospect of longer growing seasons, substantially warmer winters with little change in the summers (except, perhaps, a lengthening), and a greener world, the foreshadows of which are growing in a laboratory in Arizona.

Footnote: Forecasting Disaster by Ignoring Night and Day

In a 1990 paper in the *Journal of Geophysical Research*, David Rind, James Hansen, and their coworkers on the NASA GCM wrote:

> The likelihood of future drought is studied on the basis of (indices calculated from our climate model). Both indices show increasing drought for the United States during the next century with the effects becoming apparent in the 1990s. The model results suggest that severe drought (5 percent frequency today) will occur 50 percent of the time by the 2050s.

That is a pretty gutsy forecast, and Hansen even told *Science* in the summer of 1989 that, barring a big volcanic explosion, the *early* 1990s would see an increase in drought frequency. That forecast clock began ticking on January 1, 1990, and will expire on December 31, 1993.

Needless to say, the expectant crowd has been watching, because the United States just completed its *wettest* decade. But Rind or Hansen may turn out to be right for the wrong reason. The index that they used to calculate drought—called the Palmer Drought Severity Index (PDSI)—is one of the most abused measures in climatology. One exercise that I subject students to is to have them build a computerized model of crop yield using either total monthly rainfall or the PDSI as driving variables. Invariably, the PDSI, which drives many appeals for public funding of agricultural disaster relief, is a *poorer* predictor of crop yield than is total observed rainfall.

There are several reasons for that, not the least of which is that the PDSI is calculated by aggregating mean daily temperature, defined as the high and low divided by two. Yet a clear day and a cloudy day can often have exactly the same mean temperature but radically different evaporation: the cloudy day has relatively cool daytime readings (low evaporation) followed by a warm night, and the clear day is blazing hot (high evaporation) and is followed by a cool night. The cloudy day does not harm crops, but the clear day shrivels corn.

The PDSI calculation uses only mean temperature. If that rises, as it does in all of the GCMs, the PDSI's inability to discriminate between day and night will make subsequent calculations useless if warming occurs primarily at night. That is what would happen if clouds increased, thereby warming the nights rather than the days. The PDSI will not notice it and will scream "drought" while bins are bursting with corn.

The spring 1991 explosions of Mt. Pinatubo, the large Philippine volcano, have shortened the test of Hansen's forecast. Most research suggests that cooling from a major volcano tends to maximize in the second year after the blast, so we are going to be charitable and cut off the bet at eight months after the fact, or by the end of January 1992. That adjustment would allow Hansen's four-year *Science* forecast (January 1990 through December 1993) to run for 52 percent of its total period. Inspection of our plot of PDSI history in Figure 7.6 shows how well the forecast made out through January 1992.

Is the New Vision Greener?

Most of the plants that we live with and depend on evolved in an atmosphere with a considerably higher CO_2 concentration than exists today, and scientists have long known that raising that concentration stimulates plants' growth as long as they have other adequate nutrients. As of 1992 there were hundreds of papers in the refereed scientific literature demonstrating that phenomenon.

Enhancing the CO_2 concentration has another surprising effect: plants begin to use moisture more efficiently. Tiny pores on leaf

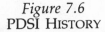

Figure 7.6
PDSI HISTORY

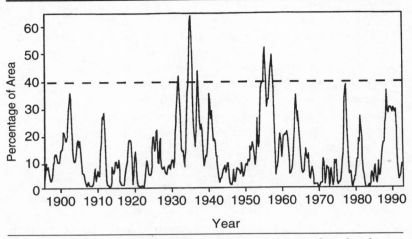

NOTE: The history shows no pattern whatsoever that can be related to an enhanced greenhouse. NASA's James Hansen said in 1989 that the early 1990s would show an increased drought frequency. We have placed a dashed line where values would have to have been for that forecast to have been correct. They were not even close.

surfaces (stomata), through which gases and water exchange, become more resistant to the flow of water and often become fewer in number. Thus, we can expect that plants of the 21st century, on average, will grow more rapidly per unit of moisture consumed. If the climate of that era is dominated by night warming—which will lengthen the growing season without substantially increasing the natural evaporation rate—we may expect a profound "greening" of the planet.

Most plants, many of which evolved during the past 100 million years or so, have had a difficult time adapting to the last 5 million years, during which the atmosphere has been most impoverished in CO_2. The best way to put plants back in harmony with their genetic material may very well be to place them in an environment more like the one in which that material evolved. We are currently so impoverished in CO_2 that Hugh Ellsaesser (1990) of Lawrence Livermore National Laboratory noted that at the height of the last ice age, the earth's atmosphere was within 100 ppm of being unable to support plant life because of a lack of that essential plant nutrient. In fact, during the past 50 million years, the earth has undergone a general cooling associated with a reduction in atmospheric CO_2 that has been reversed only by human industrial activity.

The enhancement of plant growth by increased CO_2 is known as carbon dioxide enrichment, or CO_2 fertilization, of the biosphere. Has it begun in earnest? The current concentration of 357 ppm is one-third greater than the pre-industrial background of approximately 270 ppm and 25 percent greater than values estimated for the mid-19th century when the British Empire was in its prime. British ecologist Ian Woodward (1987) found that plant specimens from the British Museum collected in the mid-19th century have more stomata than members of the same species collected today. His finding verifies laboratory experiments that demonstrate that plants grown under enhanced CO_2 have fewer stomata than those grown under ambient concentrations.

Several "real world" surveys also indicate that vegetative growth is accelerating as a result of the increase in CO_2. The effect has been noted by botanist V. C. LaMarche (LaMarche et al. 1984) in tree-ring studies of long-lived bristlecone pine trees in the western United States, in a virgin forest plot being carefully monitored by the U.S. Department of Energy at Oak Ridge National Laboratory, and in the northern forests of Scandinavia. In all those studies the growth responses have been dramatic and appear to be unexplainable by any conventional weather variable.

What may be most intriguing is that growth enhancements are occurring when forest growth is ostensibly being retarded by acid precipitation. Acid rain is thought to be a by-product of anthropogenerated sulfur dioxide emissions that accompany the burning of fossil fuel. Thus, while the same mechanism that causes acid rain may be preventing a disastrous warming, by increasing cloudiness, the same gas that should cause warming (CO_2) could be enhancing growth that will more than compensate for any putative losses of forests to acid rain. That is precisely what Pekka Kauppi (Kauppi et al. 1992) found in a study of European forests published in *Science*.

Crop Yield and the New Vision

If the daily temperature range (difference between high and low) drops, which is what I hypothesize should happen with an increase in cloudiness, could there be an increase in crop yields? My doctoral student David Stooksbury developed a series of statistical models to test that proposition, and the results were first published in the British journal *New Scientist* in 1991. That work pertained to corn yields in the southeastern United States, which is an intensive agricultural and silvicultural region. We caution against generalizing Stooksbury's studies to the world, which would be inappropriate until larger area studies are done.

Stooksbury first removed increasing trends in yield due to the effects of improvement in agricultural technology. After allowing for mean differences in yields between places, those trends account for almost 70 percent of the year-to-year variability in crop yield. Weather variation accounts for only around 15 percent if considered simultaneously with technology, which itself should make us immediately question glib projections of a climate-induced agricultural apocalypse. However, if we remove the technological trend, weather explains about 50 percent of the "detrended" yield, after allowing for the massive increases in yield resulting from technology.

We studied 40 years of corn yield data from the southeastern United States, relating them to monthly mean minimum and maximum temperature and to total precipitation for June through September. In that region, there is usually enough midsummer rain to promote good yields. In fact, there is so little variation that maximum and minimum temperatures are much more related to final yield than is July rainfall. The higher the maximum temperature, the lower the yield, and the higher the minimum temperature, the higher the yield.

Therefore, *as the daily temperature range declines, yields go up*. The change in yield for each decline of 1°C (1.8°F) in the mean temperature range is 1¼ bushels per acre. As detailed later, that change in the daily temperature range is already occurring. Over the U.S. southeast, there has been a reduction in the difference between high and low temperatures of approximately 0.6°C (1.0°F). That alteration, which is consistent with a benign greenhouse scenario, has resulted in a net increase in revenue of several million dollars per year for the region.

Other evidence consistent with vegetative enhancement is beginning to appear: Brush surveys in the state of Texas indicate a change in the area occupied by woody plants from 88 million acres in 1963 to 106 million in 1983. Sherwood Idso (1989) quotes a 1988 study describing "a dramatic invasion of the (Patagonian) steppe by tree species . . . as well as many species of tall shrubs." That change has occurred in spite of purposeful deforestation by European settlers and without any demonstrable climate change. Those and other changes have prompted two rangeland specialists, Herman Mayeux and Hyrum Johnson (1986) of the U.S. Department of Agriculture, to comment that "an experiment assessing the consequences of increased CO_2 levels has already been conducted on a global scale, and we should consider the possibility that the recent increases in abundance of woody plants on rangelands is its most visible result."

Why all of that would occur, from an evolutionary standpoint, is something of a mystery. Is it the case that there was considerable selective pressure on plants of the past for moisture efficiency because the greenhouse-enhanced world was one in which moisture stress was greater or more frequent? That is what would occur if rainfall did not increase while daytime temperature did. But if we are now increasing cloudiness—perhaps with other industrial emissions that go into the air along with increasing CO_2—are we not making plants more water efficient even as we *do not* increase the moisture stress?

Perhaps the most spectacular implication of what enhanced CO_2 may do to plants lies in a grove of sour orange trees planted by Sherwood Idso, who is employed by both the USDA and Arizona State University. Although his doctoral training is in soil physics, Idso's diverse and prolific academic publication record will surely cause him to be remembered as one of the late 20th century's most productive environmental scientists long after his passing and that of those whose academic feathers he has ruffled.

Idso's experiment, "Sherwood's Forest," has been to plant trees in the ground (rather than in much more artificial laboratory pots, as almost all other studies have done) and to raise the concentration of CO_2 in the surrounding air by 75 percent. He has compared that group of trees with a control group grown in exactly the same conditions except that they are bathed in ambient air with its current concentration of CO_2. Plants in the enhanced CO_2 started to produce fruit a year earlier than the others, and after three years they were 2.8 times as large as their nonenhanced counterparts. That enhancement means more leaves, more stems, and more roots—more green plant. By every measure of plant vitality ranging from fruit set to the mass of the roots, the carbon dioxide–enhanced trees outperformed their "natural" counterparts.

Unfortunately, for all his hundreds of refereed scientific papers and books, including the iconoclastic *Carbon Dioxide and Global Change: The Earth in Transition* (1989), the most heavily referenced nongovernmental book ever produced on the subject, Idso has an image problem. His controversial calculation in a 1980 paper, which concluded that net warming for a doubling of CO_2 would be far less than any amount necessary to support the Popular Vision, brought on one of the most vigorous and personal prosecutions of an honest mistake in the history of climatology. Curiously, no analogous witch hunts have pursued those who have erred on the high side of global warming.

Idso never attempted to mollify his detractors by the traditional route: attendance at scientific meetings followed by drinks and dinner, with proper displays of humor and humility. In fact, there is no scientist alive with Idso's record of productivity who has not made mistakes, and yet Idso's treatment was truly special. Why did he not mollify his critics with the personal touch? He has seven children, a strong Mormon background, and an intense dislike of airplanes. But his papers keep being published in the academic trade journals, even though *Science* recently refused to send his orange tree article out for professional review, saying it was insufficiently important.

Remarkably, in a experiment far more realistic than many others to date, a scientist finds that the gas concentration that we are creating and putting into the atmosphere makes trees grow three times larger, and that is "unimportant."

8. The New Vision: Verification by the Data

As noted earlier, the Historical Climate Network (HCN) of Thomas Karl for the lower 48 states appears to be the most reliable of the land-based temperature records when judged by comparable satellite measurements (since they became available in 1979). The record shows a rise in annual mean temperature of approximately 0.8°C (1.4°F) from its beginning in 1895 to 1935. Since then and during the period in which three-quarters of the greenhouse enhancement took place, there has been a net cooling of about 0.3°C, or 0.5°F.

But what is more bedeviling for the Popular Vision lies in the details of the HCN. Mean annual *maximum* temperatures—simply all the high temperatures for each station divided by 365 days and then summed over the entire country (after adjusting for station density)—show the same early rise as overall temperatures, but since 1935, they have fallen by more than twice as much as the overall temperature.

A remarkable disparity appears when maximum and minimum temperatures are compared (see Figure 8.1). Both records are congruent to approximately 1950, rising in tandem through 1935 and then showing a slight cooling. But after 1950 the low temperatures start to rise while the highs fall. The disparity is best appreciated by subtracting the low temperatures from the highs, which gives the daily range, or difference between late night and late afternoon readings (Figure 8.2). That range shows little change through mid-century and then begins to drop dramatically.

Another desirable way to examine the HCN is to break it down into seasons, especially in light of the fact that summer daytime warming is required to bring about most of the negative aspects of global warming. Summer daytime temperatures stopped warming around 1950, concurrent with the major increase in the greenhouse gases. At the same time, the nights became balmier. That is hardly

Figure 8.1
ANNUAL AREALLY WEIGHTED NATIONAL TEMPERATURES, ADJUSTED FOR EFFECT OF URBANIZATION, JANUARY–DECEMBER

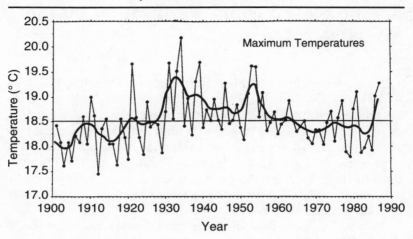

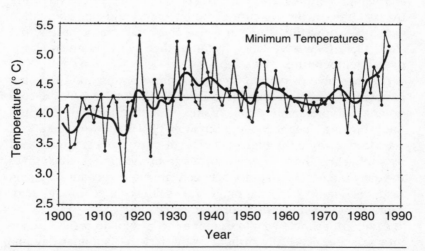

NOTE: These are maximum (day) and minimum (night) temperatures in the HCN.

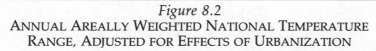

Figure 8.2
ANNUAL AREALLY WEIGHTED NATIONAL TEMPERATURE
RANGE, ADJUSTED FOR EFFECTS OF URBANIZATION

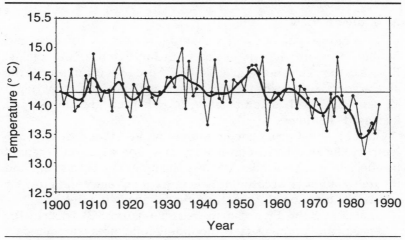

NOTE: It is quite apparent in the HCN data that a reduction in the daily temperature range began as the greenhouse enhancement became important around 1950. A likely explanation is an increase in cloudiness.

the prescription for an apocalypse; indeed, it argues much more forcefully for its opposite.

The argument that this is only a small portion of the world need not apply to this disparity. What has happened is consistent with a change in the daily radiation balance of the atmosphere (increasing cloudiness), rather than a random fluctuation that is balanced out somewhere else on the planet. There is simply no good physical reason for such a change other than some type of screening of the thermometers during the day and a blanketing of them by night. Thus, the cloud-and-greenhouse model fits.

Much was made of the lack of any overall warming trend in the U.S. record in a *Geophysical Research Letters* paper by Kirby Hanson and others. They detailed a record that is slightly different from the HCN and that includes big cities, but those cities happen to be a small fraction of the mostly rural U.S. network. No climate model that attempts to reproduce the behavior of the past 100 years has ever suggested that the United States would show a statistically

113

significant cooling trend, like the one that appears for 55 consecutive years in this record, in the face of the great increase in greenhouse gases.

A Mountain Vista

For centuries one of the retreats from the humid summers and hubbub of southern Europe has been the Pyrenees Mountains, thanks to their elevation and isolation. Those two features make the area a good checkpoint for day-versus-night warming that might be associated with a combination of enhanced greenhouse and increased cloudiness.

One of the best maintained weather stations of the past 100 years has been the Pic du Midi observatory, elevation 9,400 feet. It cannot be affected by urbanization because the mountain is so steep, and it surely cannot be accused of having trees or vegetation grow up around its thermometer, because it is far above the tree line.

What a tale the Pic's climate record (Figure 8.3) tells! In the absence of any of the usual confounding influences, summer daytime temperatures show an undeniable decline, while the nights increase in warmth. Winter days do not show a significant change, but the nights warm dramatically.

The record corroborates what shows up in Tom Karl's HCN: the daily temperature range declines in a fashion that is consistent with an increase in cloudiness and an enhanced greenhouse. Hardly apocalyptic. In fact, what comes out of the Pic record—pristine and undisturbed by any local effect, but reflecting global trends—may be the opposite: we are creating a world where winter and summer nights warm while summer days cool.

Probable Consequences

Four of the likely consequences of a cloud-mitigated greenhouse warming, all of which have to do with day and night temperatures during different seasons, could hardly be ascribed to random numbers or dumb luck.

If clouds are mitigating the warming, then their effects should vary between day and night in the annual cycle. When the days are the longest (summer), the warming should be most attenuated or perhaps even canceled entirely. When the nights are longest (winter), the combination of enhanced greenhouse and enhanced

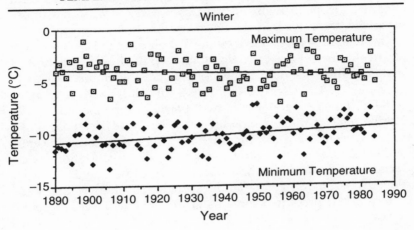

Figure 8.3
TEMPERATURES AT PIC DU MIDI OBSERVATORY

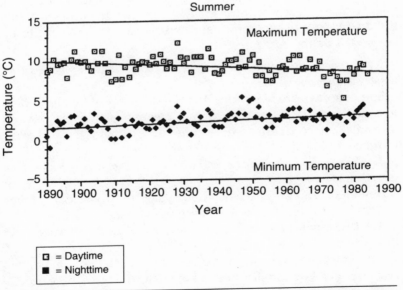

NOTE: High and low temperatures in winter (top) and summer (bottom) at the remote Pic du Midi Observatory show an obvious warming of winter nights and cooling of summer days.

clouds should create considerable warming. Parenthetically, because the Siberian Express–type of airmass forms under calm conditions in polar night or twilight, it is no surprise that those masses are the only thing in the Arctic record to show any warming.

The time of shortest days (winter) should show less mitigation of greenhouse warming by cloudiness, while the time of shortest nights (summer) should show less enhanced warming.

The U.S. HCN suggests that those difficult hypotheses may be fulfilled, but it covers only a small portion of the globe or of the Northern Hemisphere. Tom Karl and others have been busy trying to construct ersatz HCNs for other large nations, and they recently produced a record that is relatively free from urbanization for the former USSR and mainland China. In aggregate, the three HCNs now cover 42 percent of the land mass of the Northern Hemisphere.

Karl (1991) presented the data at the national level in a recent article in *Geophysical Research Letters*, and the results suggested that warming has occurred primarily at night. But a more compelling pattern emerges if the data are aggregated and weighted for the relative size of each country: each of the four hypotheses about the distribution of warming mitigated by clouds turns out to be supported by the data.

1. The cooling (daytime effect) from clouds should be pronounced on (long) summer days.

Summer days are when clouds have the greatest length of time to reflect away radiation and counter greenhouse warming. When all available records are used, summer days actually show a *cooling* trend of $-0.4°C$ ($-0.7°F$) on a 100-year scale.

2. The warming (nighttime effect) from clouds should be attenuated on (short) summer nights.

Summer nights are also when clouds have the least length of time to trap heat. Table 8.1 shows that to be the case.

3. The warming (nighttime effect) from clouds should be pronounced on (long) winter nights.

On winter nights, clouds have the greatest length of time to trap heat. Winter low temperatures show a warming of $1.8°C$ ($3.2°F$) with a phenomenal $3.8°C$ ($6.8°F$) increase in the former Soviet Union and an increase of $4.3°C$ ($7.7°F$) in mainland China. Russian winter nights give rise to the Siberian Express, so it should not be surprising that those airmasses also show some warming when they pass

Table 8.1
AREA-WEIGHTED AGGREGATE FOR THE UNITED STATES, CONTINENTAL CHINA, AND THE FORMER USSR (TEMPERATURE TRENDS [°C/100 YEARS])

Season	Mean Max. (day)	Mean Min. (night)
Winter	+0.6°	+1.8°
Summer	−0.4°	+0.4°
Annual (12 Mos)	+0.05°	+1.1°

NOTE: These are seasonal day (maximum) and night (minimum) temperature histories aggregated since records began in 1901 (U.S.), 1936 (former Soviet Union), and 1951 (mainland China) and referenced to a 100-year base. The data support each of the four hypotheses concerning the day and night breakdown as modified by cloudiness.

over Alaska, as noted by Kalkstein (1990). Curiously, no warming has been detected by the time the airmasses get to the lower 48 states, which might occur if they moved faster or if the snow cover over which they travel has expanded.

   4. The cooling (daytime effect) from clouds should be attenuated on (short) winter days.

   On winter days, clouds have the least length of time to reflect away radiation. That premise also turns out to be true, for the warming of winter days is only one-third (0.6°C or 1.1°F) that of winter nights.[1]

   On an annual basis, the observed ratio of night-to-day warming is greater than 10 to 1. Thus, it appears that four disparate hypotheses, all explainable by an increase in low-level cloudiness that is mitigating a disastrous warming, may continue to be entertained.

---

   1. All of those events are quite independent in the statistical sense. If the correlation between, say, winter temperature and that of the previous summer was not so close to zero, the success of predicting whether or not the next season will be above or below normal would certainly be greater than the currently abysmal 57 percent that characterizes most long-range forecasts, including those of the U.S. National Weather Service. We could obtain 50 percent results from forecasting above or below normal by flipping a coin. And the same lack of correlation applies, in a slightly different sense, for our day to night comparisons. In general, warm nights tend to follow warm days, but we were examining the proposition that night warming should be enhanced compared to day warming.

Clouds are the likely reason that the Popular Vision is failing in the dark.

When these results were first presented in September 1991 in an invited paper to the American Meteorological Society's Meetings on Applied Climatology and Agricultural and Forest Meteorology, one comment was, "Well, the [climate] models do not predict such a disparity, although they might have a slight night bias."

True enough, climate models—particularly the newer ones—tend to partition most of their warming into high-latitude winter, which means into the night (at the poles the night lasts for six months). But in the few attempts that have been made to use GCMs to explicitly examine the day-night breakdown in more temperate latitudes (such as the 42 percent of the Northern Hemisphere's land area that was used in this study), there is only a slight tendency for a bit more night warming.

Thus, what should be one of the simplest and most fundamental of the forecast products—the difference between night and day—appears to be off by an order of magnitude (10 to 1) in those models. That would not be so bad if the models were intended only for academic consumption, but they are the basis for policy recommendations of the United Nations. That use can be defended only if "the data don't matter."

An Exception That Proves the Rule

Although the seasonal day and night temperature histories neatly fulfill the hypotheses about increasing cloudiness in aggregate, they break down in spring over the United States (Table 8.2). Theory

Table 8.2
SEASONAL TEMPERATURE CHANGES (°C/100 YEARS), UNITED STATES ONLY, SINCE 1951

Season	Mean Max.	Mean Min.
Winter	−2.1°	−0.7°
Spring	+2.3°	+2.5°
Summer	−0.3°	+1.0°
Autumn	−1.7°	+1.3°
Annual	−0.5°	+1.0°

NOTE: This table shows the seasonal breakdown of maximum and minimum temperature changes over the United States since 1951. Spring shows the greatest warming, in apparent contravention of the cloudiness hypothesis. In fact, spring is the only season in which clouds have not increased.

predicts that winter nights should show the greatest warming, because the clouds have the longest time to hold in the warmth of the surface, and that winter days should show the least relative cooling, because clouds have the least amount of time to reflect away the sun's energy. Nonetheless, in the United States *spring* days show the greatest warming.

The spring warming seems mysterious until the seasonal history of clouds is examined. As noted earlier, Jim Angell (1990) of the U.S. Department of Commerce recently examined airport records of cloudiness across the country, comparing the period of 1950–68 to that of 1970–88. He found an overall increase of 3.5 percent between those periods. The increases were significant *in every season but spring*, which showed no change.

Another reason for a dearth of winter warming is that the eastern United States has been plagued by an increasing amount of air flow from the Arctic during much of the past four decades. In his work published in a 1992 issue of *Theoretical and Applied Climatology*, Philip J. Stenger found that, since upper air records began in 1948, the net amount of wind from the northwest had increased significantly in the eastern half of the United States, with most of the increase taking place during the 1950s and the early 1960s. As noted earlier, there was also a remarkable increase in the frequency of polar outbreaks from the late 1950s through 1989.[2]

No one really understands why (with the very recent exceptions of 1989–91) U.S. winters have tended to be so cold in the past four decades, especially as cloudiness has increased. But in spring, the only season with no change in cloudiness, both day and night temperatures increased dramatically. Thus, clouds may already have prevented several degrees of warming over continental regions. In fact, daytime temperatures in spring—the only season in which cloudiness has not increased—buck the overall U.S. annual trend by 2.8°C (5.0°F).

---

2. It is too early to tell if there has been any trend established by their absence in the eastern United States since 1989, because the region went 17 years (1942–59) without a decent one. The late James Murray-Mitchell, chief climatologist for the Department of Commerce, often lectured that a considerable portion of the decline in mean winter temperatures that took place over the United States from the 1950s through the early 1980s (when he wrote his paper) could be ascribed to some of the great Arctic outbreaks, such as Deep Freeze of January 1977.

Competing Explanations

No one really knows why the amount of northwesterly air flow into the eastern United States increased at midcentury and continued at a relatively high level into at least the mid-1980s, but there are three hypotheses. One says that the atmosphere is simply "chaotic" and changes in nonlinear (i.e., sudden) fashion between different states. Needless to say, that theory is the pet of Popular Vision and apocalypse advocates, because it is an easy stretch to say, "Well, O.K., it really hasn't warmed up very much, but the chaotic nature of the atmosphere says it will probably do so all at once." Others, more parsimonious, believe that view simply creates another ghost in the machine.

The second explanation is that there are very long, natural, and nonchaotic oscillations (of, say, 50 years in length) in Pacific Ocean currents and that those currents steer the jet stream into different, but persistent, positions. That view is supported by the repeated observation that the main center of action for jet stream changes over the entire Northern Hemisphere is in the eastern North Pacific Ocean and that some quite persistent and internally consistent patterns have arisen in that area.

The third hypothesis, which is surely the slinkiest and sexiest, is that particulate emissions are selectively cooling portions of our hemisphere by increasing cloudiness. Because the jet stream separates polar cold from tropical warmth, the concentration of those emissions in eastern North America and Eurasia suggests that the jet stream could be deflected somewhat to the south in those two regions. The result would be an increase in northwesterly air flow and an increased frequency of transport of cold polar air to the regions.

Evidence for Increasing Cloud-Enhancing Particles

Several species in the atmosphere can accomplish that trick, but they are not always available. One, sea salt, would seem to be everywhere, but ironically, not too far above the ocean's surface, it is hard to find because there is often too little upward motion to carry dried ocean spray aloft. Highly volatile dimethyl sulfide, produced by oceanic plankton, is an especially good nucleating agent, and it proliferates like wildfire if either the surface warms a bit or the plankton are stimulated by CO_2, but it is not always available either.

In fact, so-called cloud condensation nuclei (CCN) are in such short supply that the artificial stimulation of cloud development (and, it is hoped, rain) became a rewarding trade 30 years ago after Bernard Vonnegut discovered that silver iodide could create clouds where there had been none by providing a nucleus for water to coalesce about.[3]

Over the past century, human industrial activity has also been adding CCN to the atmosphere, primarily in the form of sulfate aerosol, which results from the oxidation of sulfur dioxide after the burning of fossil fuel. The concentration of sulfate aerosol is now three times the late 19th-century background, and it is continuing to go up worldwide, although emissions from the United States peaked more than 10 years ago and will drop further as a result of the 1990 Clean Air Act.

As will be seen later, sulfate aerosol is insufficient to explain the lack of warming and distribution of seasonal and day-night temperatures, but CNN are in all likelihood a partial cause.

Have people really put sulfate aerosol into the atmosphere in climatically meaningful amounts? A paper R. Mayewski (1990) and his coworkers published in *Nature* says so. Mayewski's research team drilled cores through the Greenland ice cap—a region partially downwind from the industrial regions of North America but, frankly, a little too far north to be in the main plume.

That ice field easily yields the chronology of sulfate and other aerosols, because the surface compacts some every summer, resulting in striations that can be counted backwards like tree rings. Along with each year's snow is a record of what either was incorporated in the snow or went along for the ride in the same air stream. Figure 8.4, adapted from Mayewski's (1990) paper, shows that the amount of sulfate particulate trapped in the core has increased by a factor of 2 to 3 over the past century. Nonetheless, the record is pretty uneven, in part because of variations in the amount of snowfall,

---

3. For those who have made the connection to "Ice-nine"—the condensation nucleus with the geometry that seeded a freezing of the world's water at room temperature in Kurt Vonnegut's *Cat's Cradle*—the author got the idea from his relative.

Figure 8.4
MAYEWSKI'S RECORD OF SULFATE IN ICE CORES

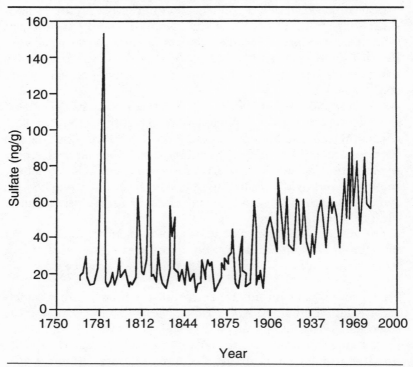

NOTE: Mayewski's (1990) ice core record of CCN-type aerosol demonstrates that it has increased by a factor of 2 to 3 in the 20th century. Further, the current load is equal to that of 1815's volcano Tambora, which caused "the year without a summer."

slight changes in prevailing winds, or the overall loading upstream in North America, which can change with the weather and the economy. Another reason for the unevenness of the record is that occasional volcanic eruptions, such as that of Alaska's Katmai in 1912, induce spikes that last for a year or so.

In fact, the largest preindustrial spike in the late 18th century was a geographic accident. It was the eruption of the volcano Laki in Iceland. One could have gotten the same type of reading by sampling a soil core around Clark Air Force Base in the Philippines in 1991 after nearby Mt. Pinatubo blew up.

But the spike in the early 19th century was real and belies a global climate anomaly. That spike was caused by the volcano Tambora, which lies completely across the planet in the Indonesian archipelago and slightly south of the equator. Tambora is the stuff of climatic legends including that of "the year without a summer" (1816) in which central North Carolina recorded June frosts and New England had snow every month.

Actually, such events normally are not that far from reality: June frosts in the mountains of western North Carolina occur today, and New Hampshire's Mt. Washington gets to within a few degrees of snow in most months.

The anomalies of 1816 were chronicled all over eastern North America and even provided a lasting memory of people since departed, such as a New England gravestone with a epitaph commemorating the deceased because he was smart enough to plant wheat (which matured in a cool environment) rather that corn (which did not) and thereby saved his family great tribulation.

Scientists have also looked at other evidence from Tambora and other great volcanoes and concluded that they probably lowered the global mean temperature between 1°C and 2°C (1.8°F and 3.6°F). University of Maryland climatologist Alan Robock has calculated that the greatest cooling effect does not appear until a year or so after the explosion; as a result, 1816, not 1815, showed the big anomalies. After two years or so of cooling from a single stratospheric blast like Tambora, the particulates begin to dissipate.

Parenthetically, Mt. Pinatubo's eruption in spring 1991 appears to have equaled or exceeded any explosion in our century, putting it in at least the class of Katmai, and probably bigger. Its cooling effect, which appears to be about half that of Tambora, should be at its maximum in 1992. Although cause and effect will be impossible to establish, October 1991 saw the greatest satellite-measured snow cover in the Northern Hemisphere since records began 25 years ago. Between September 1991 and May 1992, satellite data indicated a drop in the lower atmospheric temperature of the Northern Hemisphere of 0.8°C (1.4°F), the largest measured since the first satellite went up in 1979.

Some big volcanic eruptions do not occasion the requisite amount of cooling. Everyone thought 1982's eruption of El Chichon was going to cause big cooling, but an El Niño warming of the central

Pacific (something that has occurred every few years or so since we began measurements 100 years ago) coincided with the expected coolest year of 1983 and wiped out any signal.

At any rate, the bottom line seems clear. Because of our industrial activity, we have induced a standing crop of cloud-enhancing CCN into the atmosphere that is equal to the maximum load from Tambora, the volcano that caused the year without a summer and that reputedly cooled the planet 1°C to 2°C (1.8°F to 3.6°F).

While they may survive for a year or two in the stable stratosphere, CCN do not last for much more than a month in the lower layers before they are either rained out or just fall down. Further, their attrition rate is exponential; the largest numbers fall out first. Because so little of the lower atmosphere's air exchanges between the Northern and Southern Hemispheres each year, it follows that CCN enhancement should be primarily a Northern (industrial) Hemisphere phenomenon. Further, the rapid fallout in the first few days after emission suggests that the cloud enhancement effect should be greatest near the most concentrated source regions.

One of those regions is the eastern United States, which has a great conglomeration of coal-fired power plants in the eastern Midwest and the Ohio River Valley. Are there more or brighter clouds immediately downwind? Either one or both of those results could occur. A cloud could be made where one would normally not appear, or a standing cloud could be brightened because an increase in CCN would result in smaller cloud droplets, which would make the cloud more reflective.

Robert Cess (1989) of the State University of New York at Stony Brook provides some tantalizing evidence, even though the record is far from exhaustive and to date has been seen only at scientific meetings. He looked at a month's worth of weather satellite data from the stationary GOES orbiter and sorted them into categories including one that should indicate low-level water-vapor (stratocumulus) clouds. That species is most effective at cooling the earth's surface because it is very bright (reflecting away sunlight) and occurs in extensive sheets over large areas. The night warming effects of stratocumulus are not thought to be sufficient to outweigh their overall cooling.

Cess took a "transect" of satellite data in the midlatitudes from roughly Cape Hatteras across the Atlantic Ocean to Europe, and he examined the record to see if each data cell tended to become more reflective as the readings approached North America. Indeed, the western (North American) end reflected away approximately 8 percent more radiation than the clean eastern end, as is shown in Figure 8.5. That discovery is quite remarkable, because a change of approximately 2 percent in global reflectivity is sufficient to counter putative surface warming from a doubling of CO_2.

Cess did the same in the western Pacific CCN plume of Japan and Korea, studying a satellite transect from there to the International Date Line. Over that path of several thousand miles, reflectivity decreased by nearly the same amount.

Cess's reputation in the atmospheric science community is that of a very careful, methodical worker. It was therefore not surprising that he tried to find a "control" study that had as many features as possible, except the industrial CCN, of the two transects he had studied. There is such a place: Australia.

Although it is in the Southern Hemisphere, like Japan and the United States, Australia is in the latitude of prevailing westerly winds. It is also a pretty dusty place, and the idea that dust (rather than industry) particles might be enhancing the Northern Hemisphere plumes could be tested here.

Cess found no brightening downwind from Australia.

Is It Just CCN?

Are enhanced CCN the only factor that is channeling enhanced greenhouse warming disproportionately into the night? If so, the night-day warming ratio should be much more nearly equal to 1.0 in the Southern Hemisphere, which is so far from most of the CCN sources that almost all of the particulates would fall out of the atmosphere before arriving in Australia.

Writing in *Australian Meteorological Magazine*, P. A. Jones (1991) demonstrated that, averaged over the entire country, Australian night temperatures have been warming at a rate that is twice that of the day values, or 0.12°C (0.22°F) per decade, compared to a rise of 0.06°C (0.11°F) per decade in the maximum temperature. Further, across the country, the range in daily temperature correlated very significantly with the amount of cloudiness, which has been increasing.

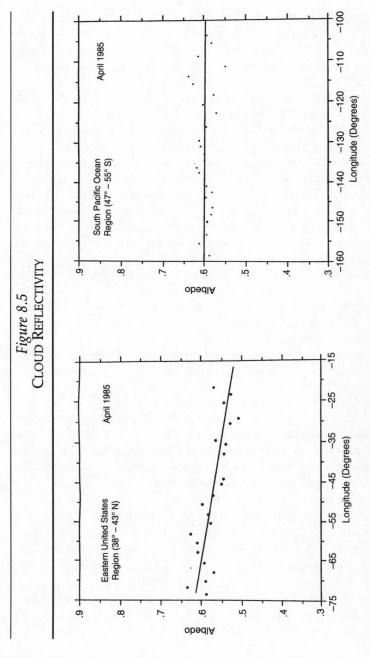

Figure 8.5
CLOUD REFLECTIVITY

NOTE: Robert Cess's (1989) two plots of cloud reflectivity show (left) from Hatteras to Europe in the sulfate plume, and (right) from Australia to Tahiti, where the plume is clean.

Analyses show . . . an inverse correlation between the diurnal temperature range and cloud cover. Significant long-term trends are evident in the data, with an increase in cloud cover and a decrease in the diurnal temperature range. There is also a slight increase in daily mean temperature.

As noted earlier, Steven Warren at the University of Washington has analyzed cloudiness over large oceanic regions of both hemispheres and found increases in both. Interestingly, although the data are sparse and very noisy, it appears that the low-level strato-cumulus clouds—which one would expect to be enhanced by low-level CCN—do not show any change around Australia. Rather, the high-level cirrus may be the culprit. Whether or not those types of clouds would make a greenhouse warming worse has been a subject of scientific debate for decades. If we believe those sparse cloud data along with the much more reliable Australian temperature record, cirrus clouds too tend to partition warming into the night.

Wasn't That Forecast?

The effect of an enhancement in greenhouse gases is to redirect downward a small portion of the earth's thermal radiation that would normally pass directly out of the lower atmosphere. That increased "downwelling" of radiation results in a slight increase in the temperature of the lowest layers of the atmosphere and the planet's surface.

At the same time, GCMs forecast that the upper layers will cool. In most of the calculations, the altitudes for which cooling is forecast are above the zone that would affect much weather. In reality, however, cooling has been observed from approximately 25,000 feet skyward, well within that zone.

One model that may have more correctly forecast the lower limit of cooling is the GFDL model shown in Figure 8.6. In that figure, which shows the change in temperature calculated for doubled CO_2, the horizontal axis is Northern Hemisphere latitude and the vertical is height. The temperature changes are for each latitude band as a whole. Consequently, the expected warming in that model at the surface for the region 80°–90° N is approximately 7°C (12.6°F), while it is around 3.5°C (6.3°F) for our latitude. It should be noted that the presentation shown in Figure 8.6, taken directly from the scientific literature, is a dramatic distortion of the areas of significant warming because it gives the impression that the land area of the highest latitudes is equal to that of lower latitudes. In reality, the highest latitude bands contain very little land area and reduce to a point at 90°, or the North Pole.

Clouds are more likely to form when surface moisture is transported to a region that is cooler than the normal "regional" temperature. That is easy to appreciate in North American winter when warmth and moisture from the Gulf of Mexico are sometimes transported northward all the way to regions where there is snow cover. The resultant fogs are legendary, and fog is nothing but a low-level cloud.

The effect can also be seen at high altitudes, again especially in the winter. When the jet stream bends way to the south, ushering in a Siberian Express or one of its slightly more benign cousins of North American origin, the temperature of the atmosphere above,

Figure 8.6
GFDL MODEL FORECAST

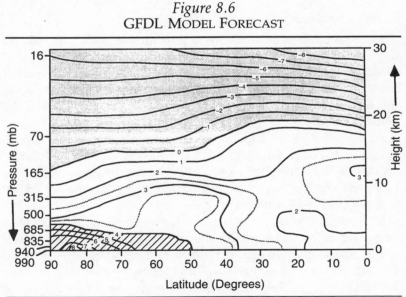

NOTE: This latitude/altitude plot from an early Princeton GCM shows how warming is forecast to be concentrated near the surface, with actual cooling aloft.

say, 25,000 feet is also very cold. Each afternoon, as jet aircraft ply their transcontinental trade, the sky blooms with clouds caused by the injection of very slight amounts of moisture from the engines into a cold upper atmosphere. Those clouds start out looking like liquid water clouds, but they soon spread into the more familiar high-elevation cirrus, or ice-crystal, clouds. Under ideal conditions, the entire sky can be covered with a thin overcast as a result of jet planes' simply injecting a small amount of heat and moisture into an unusually cold region.

Cooling has been observed at the altitudes in which such ice-crystal clouds predominate. Thus, one might expect to see an increase in observed cirrus clouds in the Southern Hemisphere. Those, too, would have the effect of warming the nights.

The indictment of the apocalypse seems to be complete. The greenhouse effect is real, but its expression at the surface of the planet (where we and everything else live) seems to be muted in the Northern Hemisphere by industrial particulates and is perhaps transformed into a benign or beneficial alteration of the atmosphere by the same industrial activity that enhanced it in the first place. In the Southern Hemisphere, even with the absence of those industrial compounds, warming appears to take place much more at night than during the day, perhaps as a result of an increase in high cirrus clouds that are a product of the greenhouse enhancement itself.

9. Measuring Climate's Impact

Any discussion of the problem of climatic change is incomplete without referring to Climate Impact Assessment (CIA). To assess how society will be affected by the greenhouse enhancement, CIA quantifies the effects of weather and climate variability on various living systems and then couples that knowledge with regional GCM output.

Of course, people are most interested in knowing the types of economic or ecological change that can be expected as a result of a changed climate. To tell them, we must first have a reliable forecast of the climate and then a reliable model to determine the resultant change in, say, forests or agriculture.

The problem is one of cascading errors (shown in Figure 9.1). We can usually relate only 50 percent of the nontechnological fluctuations in crop yield to weather and climate variability. Further, to be charitable, we will assume that we can accurately predict 25 percent of the regional changes in growing season climate as a result of the greenhouse enhancement. (At present, that percentage is wildly optimistic, inasmuch as different GCMs predict that summer rainfall in the Corn Belt will go either up or down; thus we use a more realistic 10 percent in our illustration.) We will be able to relate only $(.50 \times .25) = 13$ percent of the variation in regional crop yield to the enhanced greenhouse. And that calculation does not take into account the additional step of translating from regional agriculture to the regional economy, which is where the bottom line is.

How does CIA work? It is perhaps easiest at the outset to say how it fails to work: We do *not* have mathematical simulations that show how crop plants grow and respond to subtle changes in rainfall or temperature that we can generalize much beyond a controlled laboratory. Because agriculture is a complicated process and seemingly inconsequential variables, such as soil type, local geometry, or the farmer's peculiar personal style, get in the way of generalizing laboratory results to the real world. When trying to predict

Figure 9.1
CIA ERROR CASCADE

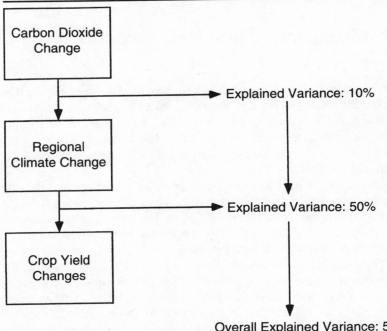

NOTE: The "cascading error" problem is at the heart of CIA. The notations on the right indicate what percentage of the behavior of each box, given the current state of knowledge, can be explained by the previous box. I use the current state of knowledge of wheat yield in the Great Plains as the example.

how agriculture will respond to climate fluctuation, CIA examines the history of crop yields and climate for approximately the past 50 years; Figure 9.2 shows a schematic CIA model.

It is first necessary to account for the fact that mean yields differ from place to place. Everything else being equal, more corn per acre comes from northern Illinois than from eastern North Dakota. Those differences result because the mean climate in some places is simply better than in others and because there is a gradation of soil quality, roughly from good to bad, from east to west through the middle of the United States.

Figure 9.2
CIA SCHEMATIC

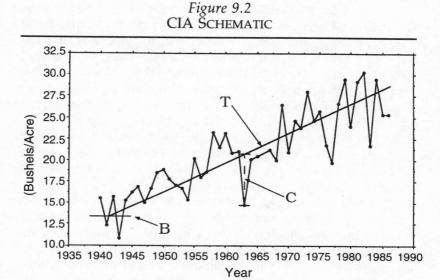

NOTE: In agriculture, CIA first examines "base level" yields (B) and their difference from place to place, next estimates the technological change (T), and then models climate (C) as the year-to-year departure from the "base" and "technology."

After CIA accounts for the mean difference between places, it attempts to account for technological changes. One of the unanswered questions about the world's food system is why the slope, or rate, of yield increase through the past half century varies so much from place to nearby place. The answer has to do with incompletely understood interactions between local economies and the fact, noted above, that not all places are created equal. Further, the more variable the weather and climate, or the lower the "everything else being equal" yield, the lower the rate of technologically related increase. Thus, farmers in more "risky" environments take fewer expensive chances.

Of course, it is the "climate term" in the overall agricultural yield equation that we are really interested in. We simply cannot simulate crop yields directly from laboratory investigations of temperature, rainfall, evaporation, and other variables, so we look to large sets of past climate-crop relationships to determine the degree to which crops respond to changes in temperature and precipitation.

133

Even specifying 12 yield predictors (6 months of mean temperature and total precipitation, for example) for the March–August crop year, along with asking the computer to objectively calculate the crop response to each, requires a simultaneous solution of 12 differential equations with 12 unknowns. The first model of this type was published in a 1920 issue of the atmospheric science journal *Monthly Weather Review* by Henry Wallace, who developed hybrid corn and was a presidential candidate.

In 1920 a 12-variable solution might have occupied an army of graduate students for a good four years. The first student to have figured out three of the parameters would probably have been awarded a Master's degree. Now much more complicated models can be flashed through an Apple computer in a couple of minutes.

But the parameters have tended to remain the same, even as some of the equation specifications have become more sophisticated. Thus, the scientific literature is still studded with this or that model whose data are on monthly mean temperature, total precipitation, or some nonlinear combination of the two.

Forest Climate Impact Assessment

Models for the other main target of CIA—large forested areas—are quite a bit different from those for agriculture, in part because trees grow for decades or centuries, while corn takes six months from planting to harvest. Further, the prime initiator of forest change is either a large-scale cataclysm, such as a fire, or a small-scale event that radically changes the immediate locality, such as an old tree's falling over or getting struck by lightning.

Those so-called gap-disturbance models account for the normal life span of various species and for what species can be expected to replace the fallen. For example, if left to its own devices, most of the U.S. southeast would turn into a pine-oak forest. The fact that the southeast is pretty much managed for timber harvest keeps it in pines.

The management aspect of forests—a growing trend worldwide—is analogous to the technological trend in CIA models for agriculture. It is ignored in gap-disturbance models. Instead, the natural replacement species is determined by a combination of monthly mean temperature, rainfall, evaporation, and radiation.

To forecast changes caused by greenhouse warming, CIA's large-area forest models simply plug in GCM forecasts for regional temperature and precipitation changes. Yet, as noted in Chapter 5, NASA GCM modeler James Hansen told the EPA in 1987 that the unreliability of regional projections from those models was the number one research problem confronting investigators of climatic change. Further, the use of mean monthly temperature (the average of the high and low) specifically ignores the fact that warming might not be equally distributed between day and night.

In summary, CIA models of large-area forests assume no technological management, knowingly use unreliable regional estimates of future climate generated by GCMs, and unrealistically divide warming between day and night. The cascading error problem becomes overwhelming because different models predict that precipitation will go up or down.

Climate Impact Assessment and the "Dumb People Scenario"

One of the nation's premier climate impact analysts, William Ribesame of the University of Colorado, presented a paper in 1991 to the American Meteorological Society in which he elucidated the concept of Climate Intelligence Failure (CIF). He used two examples: (1) the burning of Yellowstone Park and the lack of communication about how dry it really was and (2) the big flood in Salt Lake City a few years ago and the lack of communication about how wet it was.

CIFs result either when the proper information is not communicated to those in responsible positions or when the information itself is misleading. And we have got what may be the number one CIF in the entire CIA field coming out of the EPA.

A couple of years ago an EPA document on CIA hit every major daily, even though there was no official reference. The document was a "leaked" draft on the impact of climate change, complete with the familiar "do not cite or reference" stamp (therefore, I cannot cite it either). Because of global warming, it predicted that the mean yield of soybeans around, say, Charlotte, North Carolina, would be 8 percent of today's value. To come up with that number, the EPA assumed "that there would be no technological improvements in agriculture." That is about as logical as assuming black is white and saying that any subsequent painting is accurate. In fact, technology explains much more of the change in crop yield over the past 50 years than does weather, as shown in Figure 9.2.

Recently, the National Academy of Sciences prepared a document on "Adaptation to Climate Change." This is what Chairman Paul Waggoner, who has spent a lifetime studying how plants, humans, and climates interact, said in the *Washington Post* about such assumptions: "We did not foresee that people would dumbly suffer and so we did not use a 'dumb people scenario.' We assumed people would adapt. We assumed people do not have feet of clay."

10. Political Science

Famine has been estimated to have been directly responsible for 65 million American deaths in the decade 1980–89.

—Paul Ehrlich
The Progressive, April 1970

Stanford ecologist Paul Ehrlich made that 1970 forecast in a "letter to the president," dated January 1, 2000. His estimate of the number of Americans who would starve was off by 65 million. An additional 60 million *dieted* during that period. Ehrlich's statement is prototypical of a peculiar and repeating pattern that has developed around environmental issues. Its use is not limited to global warming; deforestation, species diversity, acid rain, ozone degradation, and several other issues are all presented in similar fashion.

The Apocalypse Machine

From a review of the various uses that have been made of that pattern, we can divine the rules to be followed by those who use environmental science for political ends.

1. Define the Problem As Apocalyptic

In the early 1980s acid rain was a plague that was going to wipe out most aquatic life and the nation's forests and cause an epidemic of Alzheimer's syndrome, as aluminum ions were leached into the nation's drinking water. Although the EPA wrote of an "ecological silent spring," it never arrived.

Less than a decade later, a similar threat arose from the ozone hole, which would cause millions of skin cancers and, therefore, deaths. Further, the hole would induce mass starvation by killing crops. Soon it would affect the oceanic plankton, asphyxiating it from lack of oxygen.

On the biodiversity front, which became popular a little after discovery of the ozone hole, loss of species caused by human settlement and deforestation will render humans incapable of treating diseases, which are sure to proliferate because of global warming and because of the damage to the human immune system caused by increased ultraviolet radiation leaking through the depleted ozone. Further, the ozone hole will increase the spread of AIDS.

By 1980 global warming was said to threaten the Antarctic ice sheet (which should *grow*), causing a sudden and catastrophic inundation of most of the world's major cities. Steve Schneider and Robert Chen, writing in the peer-reviewed academic journal *Annual Review of Energy*, drew a map of downtown Washington with the Washington Monument partially submerged.

Further, the planet would warm so rapidly that the few plant species left after the ravages of deforestation, the enhanced ultraviolet radiation from ozone depletion, and the leaf damage from enhanced surface ozone would be unable to migrate fast enough to more amenable climates.

2. *Present the Apocalyptic Vision As a Mainstream View: Dissenters Are Crackpots*

The apocalypse machine bases its credibility on a purported scientific consensus that provides truisms but not the whole truth. For example, it seems that every environmental interest, from the United Nations to Albert Gore, repeats that "all scientists agree that the greenhouse effect is real," a truism that is hardly profound. After the term "greenhouse effect" is defined in apocalyptic terms (Step 1) the implication is that all scientists agree that apocalypse is at hand.

In fact, there are several other things that "all scientists" do agree on that are often cited as evidence for impending doom. For example, all scientists do agree that rain is acid (i.e., its mean pH is less than 7.0). All agree that ozone in the Antarctic stratosphere declines sharply every winter, that certain species are becoming extinct, that the planet has warmed in the past 100 years, that the greenhouse effect is increasing, and that the number of people with skin cancer is going up as is the number of individuals who have Alzheimer's disease. Each one of those phenomena can be explained by a natural fluctuation of the physical environment, aging, or lifestyle choice.

3. *Play Up the Lurid Prognostications and Imagery of Doom because Apocalypse Sells Newspapers and Television Time*

Statements that the end is not near do not generate nearly as much interest as their opposite. Consequently, whatever "all scientists agree upon" is presented in as lurid a picture as possible.

A vivid example occurred in February 1991 when *USA Today* carried a front-page lead article (the lead was in red ink to emphasize warming) about the warm U.S. temperatures in 1990 and stated that 1990 was the second warmest year in a record going back to 1895. Actually, it was the sixth warmest.

In seeking background for the story, the paper contacted one scientist who suggested that they run the actual temperature history in the lower left corner of page one, which usually contains the "Snapshot USA" data factoid. The paper declined to do so because such an illustration would be "too complicated" for its readers.

I include that temperature history here as Figure 10.1. It is readily apparent that there is simply no discernible warming trend in the data. If anything, the most salient feature is the remarkable *invariance* of year-to-year temperature swings that characterize the period from the mid-1950s to the mid-1970s. The factoid that accompanied the warming article turned out to be "Measles Cases in the United

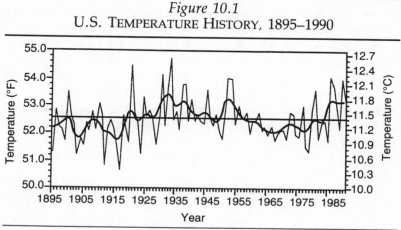

Figure 10.1
U.S. Temperature History, 1895–1990

NOTE: Because this figure obviously shows no trend, *USA Today* declined to print it as a graphic accompanying their article on the warmth of 1990.

States since 1964." Which is more arcane and which is more germane to a headline story about a possible climate apocalypse—the number of measles cases in the past 25 years or a simple line chart of the temperature history of the 20th century?

Is such uncritical coverage simply the province of common denominators like *USA Today*? Hardly. In a 10-day period in late October and early November 1991, the *Washington Post* and the *New York Times* carried the following stories:

- "Ozone Depletion Much Worse Than Previously Thought" (1/4 page);
- "Algae Bloom Possible First Sign of Global Environmental Catastrophe" (2 pages);
- "The Greenhouse Effect Is the Moral Equivalent of the Cold War" (1/4 page);
- "Soon Maine Will Look like the Great Plains, Thanks to Global Warming" (2 pages);
- "Antarctic Ozone Depletion Worst Ever" (1/8 page); and, finally,
- "Major Scientific Opposition to Global Warming Withdrawn" (1/4 page); more on this one later.

The *Post* did provide some balance with a four-column-inch article about a *Science* report that related solar variability to the Northern Hemisphere's temperature history and stated that the greenhouse signal was probably overestimated.

The algae story, originally printed in the *Boston Globe*, was especially distorted. It had been known for some time that there had been a concurrence of worldwide algae blooms. The article said that most scientists thought it was a result of global warming or an increase in CO_2, although there was a small contingent that noted a similar bloom in stratigraphic records from more than a thousand years ago. An especially lurid quote was supplied by a Scandinavian scientist who said that the bloom was the "dying canary in the coal mine."

Global sea surface temperatures, as shown earlier, have displayed little net change in the past 50 years. Current readings are a few hundredths of a degree above those of a decade ago. Yet, on scales as large as the equatorial Pacific Ocean, temperatures regularly oscillate more than 2°C to 4°C (3.6°F to 7.2°F) depending

140

on whether or not there is an El Niño event—something that has been going on ever since there has been a Pacific Ocean and a South America. It is, therefore, absurd to blame the bloom on warming. Further, had reporters done the obvious—interviewed a climatologist about the history of ocean temperatures (Reginald Newell, *the* North American expert, was a local phone call away at MIT)—they would have found that many regions where algal blooms have been caused by putative warming, such as the North Pacific, have shown an irregular cooling over most of the past 50 years.

Or maybe the change in atmospheric CO_2 concentration is the cause of the sudden bloom in the past few years. During that time the concentration of CO_2 has increased less than 10 ppm. If the algae are the response, they hardly represent the dying chirps of a coal mine canary. Indeed, the algae's response indicates that the biota of the planet are sensitive to increases in CO_2 and that minuscule changes in CO_2 will dramatically boost plant productivity. The implication is that the climate will not change much because such increased plant growth will suck most of the excess CO_2 out of the air quickly.

Are some influential reporters and news organizations fighting an environmental jihad? They are certainly under a great deal of pressure from their news sources to sign up for the jihad if they have not yet done so. *Newsweek's* Gregg Easterbrook relates the following in the July 6, 1992, issue of the *New Republic*:

> Lately [Tennessee Senator] Gore and the distinguished biologist Paul Ehrlich have ventured into dangerous territory by suggesting that journalists quietly self-censor environmental evidence that is not alarming, because such reports, in Gore's words, "undermine the effort to build a solid base of public support for the difficult actions we must soon take." Skeptical debate is supposed to be one of the strengths of liberalism; it's eerie to hear liberal environmentalists asserting that views they disagree with ought not to be heard. More important, the desire to be exempt from confronting the arguments against one's position traditionally is seen when a movement fears it is about to be discredited.

In many cases, skeptical debate simply cannot be found in the media because many have heeded Gore and Ehrlich's call for censorship and naked advocacy. Consider excerpts from the following

Minneapolis speech in May 1990 by Barbara Pyle, Environment Director for CNN.

It begins with reference to the *Global 2000* Report to President Jimmy Carter. *Global 2000*'s findings are summarized in its Executive Summary.

> If present trends continue, the world in 2000 will be more crowded, more polluted, less stable ecologically, and more vulnerable to disruption than the world we live in now. Serious stresses involving population, resources, and environment are clearly visible ahead. Despite greater material output, the world's people will be poorer in many ways than they are today. . . . Barring revolutionary advances in technology, life for most people on the earth will be more precarious in 2000 than it is now—unless the nations of the world act decisively to alter current trends.

Barbara Pyle's speech begins:

> I switched from being an objective journalist, in quotes, to an advocate . . . in July 1980, when the Carter Administration released a report called the Global 2000 Report to the President. . . . I had the very good fortune of having received that report on the same day that Ted Turner received that report, and Ted Turner hired me and gave me that report as my job description.

Later:

> We didn't become the environmental network overnight. . . .
>
> Now I have a daily show called Earth Matters, and no matter what, I have a news story on CNN period. Plus we've got Network Earth coming, and Captain Planet, which you'll all hear about in September; and I guarantee you, it will blow you out of the water. It's the first eco-cartoon. But one of the main problems about environmental reporting is that it ain't easy . . . unless [you feel you] have an axe to grind like I do. I feel I'm here on this planet to work in television, to be the little subversive person in television. I've chosen television as my form of activism. I felt that [if] I was to infiltrate anything, I'd do best to infiltrate television.

And later:

> In concluding, we the media will only have ourselves to blame if the world goes to hell in a handbasket. . . . I was

142

asked recently at another conference in Washington, chaired by [Senators] Heinz and Wirth, [do] the media have the obligation to report these issues? And I said, "Hell, no. We do not have an obligation; its a moral imperative."

4. Build Massive Financial Support

Constant newspaper repetition of environmental disaster stories, with little qualification or balance, serves as free advertising for the apocalypse machine, which is now backed by what, by several measures, must be the largest lobby in the history of the United States.

Total contributions to the six largest environmental organizations in 1988 were far in excess of $300 million from a donor base of over 10 million members. For comparison, total contributions to the two mainstream political parties in that presidential year—in which the outcome was hardly predetermined—were around $90 million from a donor base of slightly more than 2 million. There can be little doubt that the environmental political lobby is a force with much more public financial support than either political party. The trade paper *Inside Environment* listed the 1990 budgets of the major organizations, from Earth First to the Nature Conservancy, as totaling more than $600 million, or the budget of a small state not too long ago.

5. Use That Lobbying Support to Pass Economically Profound Legislation before *the Necessary Science Has Been Completed*

We have already seen two recent examples: ozone depletion and acid rain. The Montreal Protocol to reduce CFCs in the absence of firm scientific evidence of a major problem has been hailed as a victory for the environment and a blueprint for future actions, including legislation on global warming. Richard Benedick, who helped to negotiate that treaty while with the State Department, wrote in a 1992 *Scientific American* piece that global warming is analogous. He neglected the fact that in the case of global warming (as opposed to ozone depletion) the data are pointing to the opposite of the environmental harm glibly predicted by inadequate models.

Acid rain was even worse, for the comprehensive research summary was completed concurrently with the passage of the Clean Air Act. That study, known as NAPAP (for National Acid Precipitation

Assessment Program), was a few feet thick, was the world's most comprehensive document on acid rain, and concluded that the dramatic and lurid statements made when acid rain was first defined as a problem (Step 1) were greatly exaggerated. It also concluded that the prime effects in North America appeared to be confined to a minuscule percentage of eastern forest and a small sample of lakes, most of which were already acidifying from natural changes.

NAPAP involved more than a dozen federal agencies and departments; took 10 years to complete; cost nearly $600 million; received a brief hearing in the U.S. Senate with a few "attaboys" from its original sponsor, Sen. Daniel Moynihan (D-N.Y.); and was buried in a pauper's grave, while the Clean Air Act sailed on through.

Shockingly, the 435 members of the U.S. House of Representatives were apparently so busy preparing to pass the 1990 Clean Air Act (cost: $40 billion) to mitigate acid rain that they did not have time to hold a hearing to learn that the NAPAP report said acid rain was not an apocalypse after all. They were simply too busy spending other people's money (which, because of deficit financing, they did not have yet) to listen to good news that might have prevented, or reduced, some of that expenditure. One gets the impression that the need to spend money on fixing the environment is independent of the needs of the environment itself.

Since legislation was passed, the coffers of the apocalypse machine have been fattened, more environmentalists have been employed, fewer coal miners are working, general inconvenience and tax revenue have increased, and personal income has decreased. A footnote usually appears far from the front page: it was not so bad after all. Major ozone depletions will turn out to be mostly inconsequential and will be confined to high latitudes at night and in the spring. Global warming will be seen as neutral or even beneficial as long as the economic wherewithal remains to exploit it, and acid rain has already been assigned to the dustbin of apocalypse history.

6. *Invent a New One*

Spend more taxpayer dollars to repeat the process.

Media Intimidation

In 1989, when British documentarian Hilary Lawson set out to produce yet another greenhouse apocalypse show, he ran into the

144

data and characters portrayed in this book. The result was "The Greenhouse Conspiracy," a decidedly nonapocalyptic program whose central thesis was that the issue had been dramatically overplayed by a collection of special interests ranging from environmental organizations to fund-starved scientists and the media, which needed to sell a story.

While "Conspiracy" was fun to watch, it did not even mention the remarkable imbalance between day and night temperatures or the seasonal distribution of warming. Consequently, its case could have been even stronger. At any rate, after it was reviewed as "the science documentary of the year" by the *Financial Times*, American Public Broadcasting Service refused to show it, claiming it was "too one-sided." That was the same PBS that touted James Burke's "After the Warming"—complete with distorted maps and busting forecasts—as a plausible future.

PBS eventually showed "Conspiracy" on a few outlets, notably Maryland Public Television, but only after the Discovery Channel aired it twice. Discovery was skittish, too, and produced an hourlong "panel discussion" following the presentation. The moderator, Diane Rehm, controlled what was said, cutting off participants in midstatement if their words were not in agreement with her vision of the program. That panel, of course, provided official evidence of evenhandedness.

One of the panel guests was *Washington Post* columnist Jessica Tuchman Matthews, who also works for World Resources Institute. Her qualifications for speaking about global warming are not especially clear, but she holds a Ph.D. in molecular biology from Cal Tech, she was a Carter administration prodigy on the Council for Environmental Quality (which wrote *Global 2000*), and she knows Washington and international environmental happenings as well as anyone.

At any rate, while all the participants were awaiting taping, I asked the producer if she could make a copy of a slide I had of the new NCAR GCM (the same illustration can be found in Chapter 12 of this book), which showed how projected strong warming was distorted and, in actuality, confined to a very small area. Jessica rose from her chair, shaking her head, even as fellow participant Michael MacCracken of Lawrence Livermore National Laboratory dug through his slide set looking for a good illustration. "I don't know," she intoned, "I just don't think we can have this."

I was amazed and said that I simply could not believe that she would attempt to suppress legitimate information from a refereed scientific journal. "Well," she told the producer, "anyone could bring some slide that would make their argument seem very convincing, and I just don't think that's fair." And after about 30 seconds of "I'm going to walk off the show if you show that" looks from Jessica, Discovery announced that I could not show the most recent results, published and in the open literature, from the National Center for Atmospheric Research, which is sponsored by the National Science Foundation.

The Political Climate

There is little doubt that climatology is now *the* political science, as various global warming interest groups line up to make hay out of each new paper or thermometer reading. Each group, of course, hopes that whatever the newest finding is will be the silver bullet that dispels all doubt—one side hoping for final confirmation that the apocalypse is at hand, the other hoping for the opposite. And the stakes are very, very high, including costs that equal some gross national products and the further political aspirations of Albert Gore.

1990 Temperatures

The temperatures of 1990 were especially subject to political abuse. For most people who live in and around the latitude band 30°–40° N, which includes virtually all presidential candidates and an inordinate amount of the world's industry and capital, March, in particular, was incredibly warm. In the second week, temperatures topped out at phenomenal levels, with 90°F (32°C) readings recorded at many places in the eastern United States on March 13. The core of the heat, ironically, was within 100 miles of Washington, D.C.

Senator Gore was so concerned about the warm winter temperatures and bleaching corals (more on that later) that he convened a hearing. When Phil Jones (one of the authors of the Jones and Wigley temperature history) came from England to testify, he was asked to emphasize several times the warmth of March 1990. Meanwhile, the satellite data, with much more global coverage and a presumably enhanced accuracy, did not see 1990 as particularly warm at all. In fact, although March 1990 was extremely warm in

146

our latitudes, our opposites in the Southern Hemisphere were not overheated, as is made readily apparent by a comparison of Northern and Southern Hemisphere satellite data in Chapter 5. Why was that not brought out at the hearing? At latitudes 30°–40° S there was cooling, but only fish live there, and they do not vote.

Fire in Kuwait

Perhaps no event in recent history so demonstrates the exploitation of climatology for political purposes as the 1991 oil-well fires in Kuwait.

There is no doubt that the Gulf War inflicted great environmental damage on Kuwait, caused especially by the blackening of the desert surface and the hydrocarbon precipitates that will dramatically affect the local climate and chemistry when they bake in the sun. But instead of reading about that very real tragedy, we heard tired arguments by political scientists about how the fires were going to affect global climate, or at least the Indian monsoon, causing (according to at least one CNN report) millions of people to starve. Then there was the "Today Show" anchor who on the first day of the war announced that the fires would cause the earth's temperature to rise 10 degrees because of—what else—the greenhouse effect.

In the Gulf War, once again, political agendas were tuned to abuse climatology for their own purposes, and the media dutifully reported the nonscience. The first shot in this venue was fired some two months before the raid on Baghdad, when Abdullah Toukan, a nuclear physicist who serves as an adviser to Jordan's King Hussein, told a Geneva conference that the burning of the oil fields would accelerate global warming. At least he knew what riles up folks in the United States. In reality, the subsequent fires produced only a 10 percent increase over the usual daily industrial burn.

The global coolers weighed in next. John Cox, an engineer who works for Britain's Campaign for Nuclear Disarmament, said that the burning would disrupt the heat-driven Indian monsoon. Carl Sagan, American champion of the scientifically suspect but politically correct "nuclear winter," predicted global cooling because the plumes would "self-loft" into the stratosphere. University of Virginia's S. Fred Singer, debating Sagan on "Nightline," then allowed that "where there's smoke, there's Sagan." And on the network news, Germany's Paul Crutzen called for 10 to 15 degrees of cooling.[1]

All of the contentions—for major warming or cooling on continental or global scales—were based on incomplete models and unrealistic assumptions that ran counter to observations. It was quite obvious from the early photographs that the smoke plumes were remaining very low—something a climatologist (but maybe not an engineer or a nuclear physicist) would take for granted in a desert environment at that latitude. That region is desert because of lack of rain; that lack exists because the required upward atmospheric motion is inhibited by forces that are truly global in origin. If there were a tad of upward motion, vegetation would blossom, for the atmosphere itself is actually loaded with moisture. Kuwait has some of the world's highest dew point temperatures, which are an excellent measure of the total amount of water vapor in the air.

The real ecological damage went unreported: terrible damage to the desert ecosystem. The scientific disaster went unnoticed—more pronouncements of gloom and doom, made by those who must have known better, lessening the little remaining credibility of our most political of sciences.

Getting the Message Out

An examination of the data on global warming easily leads to the conclusion that forecasts of dramatic and deleterious warming are at best highly unreliable, and, to fend off critics, global warming rhetoric is plagued with the subjunctive.

Consider the recent Sierra Club "1-800-TOO-WARM" ads. (That is the number you are supposed to call after viewing television commercials featuring Meryl Streep, William Shatner, and other important people.) The Sierra Club knows full well that if it said things that were patently untrue, some columnist or pundit might complain enough to get the ads removed, or at least paid for. (The ads currently run for free because they are "public service"; the club's annual budget is a mere $42.5 million.) Anyway, to avoid

1. At a "town meeting" sponsored by the Quadrennial Ozone Symposium in Charlottesville, Virginia, in June 1992, Crutzen was asked about reports that eruptions or volcanoes, as well as chlorofluorocarbons, could result in stratospheric ozone depletion. Rather than comment on the voluminous scientific literature that indicates that possibility, he replied simply that anyone who would spread such a story was an "environmental terrorist."

those problems, almost all of the verbs in the text are subjunctive or conditional. Consider the following excerpts:

1. "America's heartland *might* have to live with temperatures over 90° [F] for almost one-third of each year."
2. "Chicago *could* expect two months of 90° [F] temperatures each year."
3. "*If* the weather in Dallas goes over 100° [F] for two and a half months a year, imagine what *could* happen in Phoenix."
4. "America's heartland *could* have trouble growing the corn, wheat, and oats that we need. The same is true for California and Florida."
5. "Drinking water *could* become a problem."

Unfortunately for those ads, all climate modelers agree that we cannot reliably use their projections for any regional climate estimations (e.g., for America's heartland). Using cities such as Chicago or Dallas for point estimates is not even science, especially inasmuch as the correlation between models for regional estimation of summer rainfall is *zero*. Further, according to research by Robert Balling (1990) of Arizona State University, extreme high temperatures in general bear little relation to mean temperatures, and it is the latter that climate models predict.

In fact, the following five statements, based on the output of GCMs, are just as accurate as the Sierra Club's.

1. The growing season in America's heartland could lengthen by up to three months.
2. Chicago currently has more than seven weeks of below-freezing temperatures. In the next century, that could be down to three.
3. If there were no more below-zero readings in Chicago, imagine what could happen in International Falls!
4. More than 800 laboratory experiments demonstrate that carbon dioxide enhances plant growth! Fruits and nuts will proliferate in California!
5. All of our climate projections say that rainfall should increase globally. We *may* have more water available to feed a growing population!

Would a television station run those predictions as a public service?

Show Trials

Although there is a wide disparity between climate projections supporting the Popular Vision of apocalypse and the observed data, differences between modeled projections and empirical data are the norm in science. A model (such as the General Circulation Model) is nothing but a series of calculations (some based purely on theory, others modified by subjectivity) designed to mimic and explore the behavior of a system. A model is a hypothesis, and science normally uses observations or data to approve, reject, or modify hypotheses. Resolution of such discrepancies is the most traveled road to scientific progress.

In the case of global warming, that normal (and constructive) tension between "data-driven" and "model-driven" scientists was interfered with by a political process eager to use the threat of dire global warming to drive policy. With its total annual expenditures in the hundreds of millions of dollars, the Apocalypse Machine has a lot of mouths to feed, and there are strong institutional pressures to keep up the image of impending doom in order to keep the cash flowing. That machine needs global warming like Democrats need a recession.

It is natural for the contributors to the Apocalypse Machine to want to see a product delivered, usually legislation. To achieve that, a political process must be initiated that requires leadership of some elected official, and in the case of global warming, the right person for the job has turned out to be Senator Gore.

Virtually every observer of Gore comes away with the following impression: Gore is bright, telegenic but a little stiff, supremely egotistical, driven, a man on a mission to save the planet, and a man who wants to be president in the worst way. "I make no bones about the fact that I want to be president," Gore told the *Washington Post* in a 100-column-inch profile on May 28, 1992. "A minute later he repeats his ambition. Then he mentions it a third time," the article relates.

Gore realizes that the normal Democratic coalition of special interests is simply not strong enough to win the White House in the absence of a major Republican scandal or terrible economic news, Watergate being about the only plausible explanation for Jimmy Carter. Further, he knows that the Democratic party has been bleeding young, white, educated thirty- and forty-somethings for a couple of decades now, and it is precisely that group that is in its most influential and financially productive years. Their party of choice is going to receive considerable financial support.

Poll after poll has demonstrated that one of the prime concerns of that group is loosely called "the environment" and that members of that group are major contributors to the Apocalypse Machine. Further, Gore genuinely shares their long-held feelings about the environment. In his 1989 stock speech he was fond of saying that he first became aware of the problem of global warming when he took a course from Roger Revelle at Harvard and that he was proud to invite Revelle to be a witness at the first congressional hearing that he ever held.

Some people saw evidence of considerable hubris when, as a freshman senator, Gore announced a presidential bid in 1988 and won a substantial number of delegates before his campaign ground to a halt in New York. Far from being hubris, that was an exploratory campaign he knew he could not win. His original speech, "This Is about the Future," was perfectly true, for he was running for either 1992 or 1996. Bush's perceived popularity in 1991 probably kept Gore from entering the 1992 presidential race, a decision he may have regretted by the summer of 1992, when Bush's popularity reached a low ebb and "none of the above" (Ross Perot) blew into the lead in states such as California, which is rich in delegates and green, too. Now that Gore has been named as Clinton's running mate, it is clear that his presidential aspirations are still in place.

In retrospect, Gore's 1988 foray was quite interesting. As a border state Democrat, almost all of his primary victories were in states that bordered or included the 35th parallel. One glaring exception was green Oregon, where he defeated considerable opposition, handily, by emphasizing the environment, the greenhouse effect, and the future. He had found the key, and unlike so many who seek an issue, he passionately believes in the Apocalypse. As he wrote in the *New Republic:*

> "Evil" and "Good" are terms not used frequently by politicians. Yet I do not see how this problem can be solved without reference to spiritual values.

Although the first sentence is clearly in error, he speaks quite clearly. Those who do not agree must be somewhat less than good. The second sentence would also disturb Thomas Jefferson, who argued for reason and the Enlightenment as bases for politics. In fact, sometimes Gore's passion, so evident, steps over the line of good taste or political sense. In March 1989 he wrote a piece for the *New York Times* in which he likened those who did not share his beliefs about global warming to those who, 50 years ago, ignored the signs of the coming Holocaust.

Now, if a scientist states publicly that global warming may not be occurring in accord with the Popular Vision, he can expect a personal attack, signed by the senator and delivered by him to a nationally prominent newspaper. Gore shares his penchant for publicly attacking the motives and credibility of scientists with Colorado's senator, Tim Wirth, also a Democrat, whose pieces in both the *Times* (about MIT climatologist and National Academy of Sciences member Richard Lindzen) and the *Washington Post* (stating there is no valid scientific objection to the Popular Vision) are beyond reply.

The peculiarity that has arisen is that the data-driven scientists— those whose research attempts to determine the validity of the Popular Vision hypothesis—are somehow painted as "conservative" or "right wing," no matter how they vote, because the Democratic party has commandeered global warming as an issue with which to break its recent habit of losing national elections. Therefore, the purveyors of environmental doom and gloom are "progressive" elements of the Democratic party.

As a result, in congressional hearings, the "data people" are usually assigned to the minority and testify singly against as many as 10 opposition witnesses in a decidedly unfriendly climate. In February 1991, Gore simply refused to acknowledge the existence of one of the data-driven scientists in a panel that he was introducing. In fact, the congressional hearing has become the prime vehicle for "show trials" of scientists who do not buy the Popular Vision.

The main bogeyman in Gore's presidential aspiration anxiety closet is Richard Lindzen of the Massachusetts Institute of Technology. Lindzen, like so many of the data-oriented scientists, spends days and nights wondering why the atmosphere has warmed so little in the face of an effective 50 percent change in its CO_2 concentration. One of many hypotheses that he has been exploring is that water vapor at high altitudes might have changed as the greenhouse enhanced, which would have canceled a great deal of prospective warming, and might explain why there has been so little.

It turns out that one of the mechanisms that Lindzen had explored did not stand the test of the data. Such testing, of course, is the way of science. When Lindzen mentioned that in his October 7, 1991, testimony before Gore's committee, it was leaked before publication to Phil Shabecoff (formerly of the *New York Times*), the executive publisher of *Greenwire*, a global warming and environmental newsletter that you can subscribe to for a not inconsiderable fee. Shabecoff then announced at a Quarterly Greenwire Briefing (he wanted to show how he could get the inside stuff before anyone else) that Richard Lindzen had withdrawn his scientific objections to global warming. Shabecoff's reputation as the *Times*'s beat man on global warming was controversial, and in public he seemed much more deferential to those promoting the Popular Vision and somewhat less civil to the data-driven scientists. He was eventually taken off the story.

Gore's staff then enthusiastically told R. J. Smith of Washington's Cato Institute that "yes, Lindzen has recanted," a word traditionally more associated with the Spanish Inquisition than with the normal process of science. Tom Wicker, who has expressed glowing support for the senator's past (and future, meaning 1996 or 2000) presidential aspirations, picked up on Shabecoff's revelation and on October 24, 1992, published a piece called "Time for Action," which was couched in much softer language, but his point was obvious.

> Richard Lindzen, (followed by his academic pedigree . . .) told a group of scientists [he was testifying before a Senate committee] that he had "withdrawn" the complex hypothesis that was an important element of his skepticism.
>
> Dr. Lindzen's views carried considerable weight in the political controversy about global warming, having been cited frequently by the Bush White House in its support of its go-slow policy. Senator Albert Gore of Tennessee, who convened the scientific group, said that while Dr. Lindzen remained skeptical on general grounds, he had abandoned the specific theory he had previously advanced.

Fair enough, but conspicuously absent is "the rest of the story." Here are the relevant parts of Lindzen's November 30, 1991, letter to the editor of the *New York Times* in which he attempted to clarify the situation. His letter was finally printed a month after the original submission and only after this original text was changed, under pain of nonpublication.

> Tom Wicker . . . refers to my having "withdrawn the complex hypothesis that was an important element of [my] skepticism." His reference is profoundly misleading and completely out of context. His conclusion that my "changed position" seems to "justify a heightened urgency in responding to threatening atmospheric developments" is totally unwarranted. . . . Two years ago, I offered as an example of a mechanism that could be consistent with the hypothesis (of a greatly diminished greenhouse warming) the fact that in a warmer environment, deep clouds might rise to greater heights and deliver less water vapor to the environment. While this mechanism is qualitatively correct, my colleagues and I found early this year that the mechanism was quantitatively inadequate. This did not (and does not) overly concern us since there are other mechanisms that are potentially more effective.
>
> Those seemingly arcane matters were not, however, at the heart of my testimony before Senator Gore. The main points I made in my testimony were [I omit the first two for brevity]: (3) In current models, the behavior of water vapor in this altitude range is largely determined by identifiable computational errors; and (4) even if these computational errors were corrected, we do not adequately know the physical processes. . . . It is my personal feeling that computational errors hardly constitute a sound basis for economically profound policy. This may be considered by some to be a political position. I hope, however, that I may be forgiven, as a scientist, for feeling that the alternative politics, which lead Mr. Wicker to urge allegiance to the results of computational errors, to be both peculiar and deeply worrisome.

Clearly, the statement by Gore's office claiming "recantation" is hardly warranted, and to say that Lindzen's statement constitutes a go-ahead signal for what he calls an "economically profound policy" is purposefully misleading. More disturbing is the question that history will ask: "Why, in the technologically sophisticated late 20th century, was it necessary to do that to science and to scientists?" Is presidential aspiration really worth breaking the credibility of a highly esteemed member of the National Academy of Sciences?

Revelle's Last Testimony

There is little doubt that Roger Revelle, one of the country's brightest scientific lights for a half century, was the person most responsible for our awareness of the enhanced greenhouse effect. As director of Scripps Institute of Oceanography, he pioneered research on the global carbon cycle. While on the Harvard faculty, Revelle taught a young Albert Gore, Jr., how the increasing concentrations of CO_2 could have global climatic consequences. In his last decade of life, as president of the American Association for the Advancement of Science, he planted the seeds that made the phrase "global change" the catchword of the nation's scientific establishment and the greatest reprogrammer of nonbiomedical research dollars since Sputnik. He saw the policy wheel begin to turn; finally, global change became so compelling that, in the last years of his life, it was being touted as the vehicle to produce, even require, "a fundamental restructuring of the world's economy."

And in his last paper, published in 1991 in *Cosmos* along with papers by Chuncey Starr and S. Fred Singer, he said the science just was not there. Global warming, he wrote, is a compelling problem, but the data simply do not support the lurid scenarios so popular now. Clearly, there are profound scientific problems that should preclude any precipitous economic action. Further, there is very little scientific agreement on the severity of the problem itself. It is better to understand what we face before we create great economic reordering: the coauthored paper finished with this sentence: "We can sum up our conclusions in a simple message: *The scientific base for a* [disastrous] *greenhouse warming is too uncertain to justify drastic action at this time*" (italics in original).

That was the last scientific testimony of the Grand Old Man of global change, the man who published hundreds of papers on the subject and who provided the first congressional testimony on global warming. As a tribute, should it not at least be entered into the *Congressional Record* by his devoted student, Senator Gore?

Blue Coral

Another apocalypse show was held in October 1990, when Senator Gore conducted a hearing about the infamous bleaching of coral in the Gulf of Mexico and the Caribbean Ocean. Corals, which have been on the planet for hundreds of millions of years, are an association between animal and plant, wherein green algae (*zooxanthallae*) assist in the digestive and hygienic processes of the animal (*anthozoa*). In combination, the two give corals their characteristic attractive colors; if the algae die, the coral appears whitened. If the algae somehow do not regenerate, the coral itself will ultimately die.

In the late 1980s a sudden, unexplained, and widespread bleaching of coral took place in the Gulf of Mexico and the Caribbean Ocean. Environmentalists instantly assumed it was caused by global warming, and Senator Gore took his now familiar path to determining scientific truth: he held a congressional hearing and carefully selected those who testified. Congressional hearings about the environment have become a peculiar combination of science and political theater. In Congress, the majority Democrats, who feel that somehow environmental apocalypse is consistent with their "progressive" views, are under no obligation to allow the minority Republicans any countering witnesses. The Democrats often concede to one, who then is fortunate enough to participate in a procedure that is reminiscent of the Clarence Thomas hearing.

No minority witnesses were allowed at Gore's coral show, and witness after witness expounded that warming would do awful things to coral and, yes, those things were happening before our very eyes. A satellite data record (which began in 1980) contained some very warm temperatures, but that record describes only the top few millimeters of the ocean—not the "mixed layer" of 100 meters or so that is inhabited by corals. The sensor was not the same as the one used by Spencer and Christy to measure the temperature of the low layers of the atmosphere. No one at the hearing mentioned that the political target—people who can afford Caribbean vacations and who scuba their way around the reefs (where touching often causes coral death)—is composed of many former Democrats and that winning them back to the party could provide a lot more support at election time.

Reginald Newell of MIT was not invited, for if he had been he would have testified that there had been no net change in sea surface temperature in the Northern Hemisphere in the past 50 years. Further, he would have noted that the rise in SST in the early 20th century—long before the greenhouse enhancement was important—was of much greater amplitude and length than anything seen since. Yet there are no records of a massive coral catastrophe.

Joe Elms of the Department of Commerce's National Climatic Data Center, who was similarly not invited, has carefully analyzed mixed-layer SSTs over the Caribbean and the gulf since 1950. He has found no overall warming trend whatsoever (see Figures 10.2 and 10.3).[2]

The argument has often been made that changes in mean temperature might not be all that important in comparison with what happens at the warmest or coldest time of the year. For example, the lack of a daytime summer warming signal and the appearance of a strong one during winter nights are hardly apocalyptic specters, but the opposite could be important. Elms reasoned that, even if the mean SSTs were not changing, he should investigate the behavior of the warmest month of each year. He did and found that the mean temperatures of the warmest months display a significant *fall* (see Figure 10.4). Thus, the data are saying the opposite of what they should to support the coral bleaching by global warming hypothesis.

There is another interesting spinoff: everything else being equal, it is the warmth of the tropical oceans during the warmest time of the year that provides energy for tropical storms and hurricanes. In *Meteorology and Atmospheric Physics*, Sherwood Idso, Robert Balling, and Randall Cerveny (1990) recently demonstrated that there are more storms in warm years, but that they tend to be relatively weak.

It seems logical that a decrease in the temperature at the warmest time of the year in a hurricane spa like the Gulf of Mexico might result in a decrease in the frequency of such storms. Curiously, some marine ecologists, such as Jay Zieman of the University of Virginia, are now citing the anomalous *lack* of recent hurricanes as the reason for a major and sudden dieback of sea grass beds south

2. There was a warming of a tenth of a degree or so in the latter half of the 1980s, but that type of minuscule temperature excursion is common in coral-dominated latitudes. In fact, mid-Pacific El Niño temperature swings that occur over some of the oldest reefs on earth can exceed $\pm 2.0°C$ in just two years. That oscillation is some 40 times greater than any found by Elms, and there have been more than two dozen El Niños in the past 100 years.

Figure 10.2
ELMS'S GRID BOX

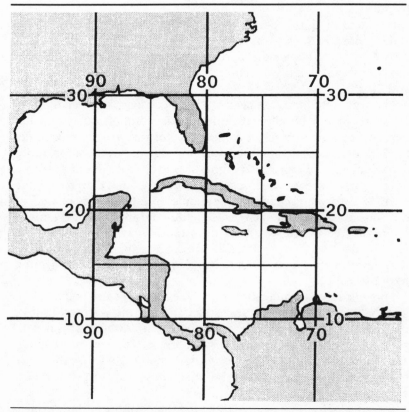

NOTE: Grid boxes used by NOAA scientist Joe Elms to measure sea surface temperature (SST).

of the Everglades, the very region in which the bleached reefs are found. Rumor had it that the sea grass dieback was caused by global warming, too.

Congressional hearings on global warming rarely reflect the honest and evenhanded debate that makes science worth doing. Rather, by judicious selection of its witnesses, the majority is usually able to "prove" whatever it wants and to make the minority position look as if it were the opinion of a scientific lone wolf. Thus, "correct" science is determined by what is *politically* correct.

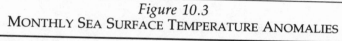

Figure 10.3
MONTHLY SEA SURFACE TEMPERATURE ANOMALIES

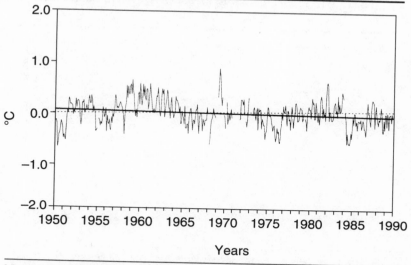

NOTE: Departures from the 1950–90 average for the region shown in Figure 10.2. There is no evidence of any recent or rapid warming (Elms and Quayle 1991).

There are some rare exceptions in which congressional orchestration of science fails. In May 1992 the House Subcommittee on the Environment met, ostensibly to consider whether or not there was a need for more research on global climate change. That meeting had the potential to put the doomsayers at a disadvantage, because any statement to the effect that more research was needed implied that their forecasts were not adequately based.

The climax of congressional hearings usually occurs when the committee chairman polls the scientific panel as to whether or not a certain piece of legislation should be supported. It is almost always expected that the vote will be four to one in favor of whatever the chair wants, as that reflects the ratio of "majority" to "minority" scientists (when a minority witness is allowed). In this case, subcommittee chairman James Scheuer (D-N.Y.) asked the scientific witnesses whether, if they were members of the House of Representatives, they would vote for a bill sponsored by Rep. Henry Waxman

Figure 10.4
SEA SURFACE TEMPERATURE ANOMALIES FOR THE MONTH WITH THE HIGHEST MEAN MONTHLY TEMPERATURE IN EACH YEAR

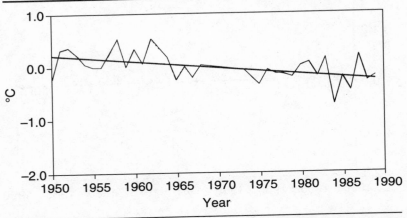

NOTE: If hot temperatures are responsible for coral bleaching, the trend, indicated by the straight line, should be increasing significantly. Instead, it is decreasing.

(D-Calif.) that would mandate that the United States stabilize its CO_2 emissions at 1990 levels to fight global warming. After one panelist cracked that he wondered if he would "get a checkbook" as a member of Congress (a reference to the notorious House check-bouncing scandal), a majority of the scientific panel failed to support the bill.

In large part, that result occurred because the nonapocalyptic viewpoint of climatic change was so convincing.

> On the one hand, we are ethically bound to the scientific method, in effect promising to tell the truth, the whole truth, and nothing but . . . which means that we must include all the doubts, caveats, ifs, and buts.
> On the other hand, we are not just scientists, but human beings as well. And like most people we'd like to see the world a better place, which in this context translates into our working to reduce the risk of potentially disastrous climatic change. To do that we have to get some broad-based support, to capture the public's imagination. That, of

course, entails getting loads of media coverage. So we have to offer up scary scenarios, make simplified, dramatic statements, and make little mention of any doubts we might have. This "double ethical bind" that we frequently find ourselves in cannot be solved by any formula. Each of us has to decide what the right balance is between *being effective and being honest* [emphasis added]. I hope that means being both.

—Stephen Schneider
National Center for Atmospheric Research
in *Discover Magazine*, October 1989

11. Competing Apocalypses: Global Warming, Ozone Depletion, and Acid Rain

We have already established that one of the reasons that warming has been muted and been occurring increasingly during the night (therefore, more into winter) is the confounding effect of human aerosol emissions, including sulfates that are involved in the production of acid rain. Thus, the same compound responsible for one apocalyptic environmental threat appears to cancel another.

In the same way, the compound chiefly responsible for the threat of global warming—carbon dioxide—may largely compensate for acid rain's putative ravage of forest ecosystems. As noted earlier, most plants that we live with and depend on evolved in an atmosphere with a considerably higher CO_2 concentration than exists today, and scientists have long known that raising that concentration stimulates plant growth.

There is another intriguing interaction between the biota and climate that has only recently been explored by Mark Schwartz of San Francisco State University: plants themselves modify the climate and make it more favorable to their growth. Schwartz has been studying the "green wave" that sweeps northward in the Northern Hemisphere every spring as leaves appear on deciduous plants. He has found that the transition of the earth's surface from bare to leaf shaded dramatically reduces the seasonal rise in daily high temperature. Further, the plant canopy is very effective at moistening the local atmosphere; therefore, night temperatures are held at higher levels. Thus, once the leaves come out, the normal daytime warming regime of the atmosphere, as the sun climbs higher through the spring sky, is reduced at the same time that the likelihood of a killing frost at night is diminished.

If we consider that the primary effect of an enhanced greenhouse warming is an increase in night and winter temperatures, it seems logical that leaves will appear on deciduous plants earlier in the

163

year. Thus, the vegetation itself will move a greenhouse warming further into the night and away from the day.

The Ozone Connection

In October 1991 UN environment director Mostafa Tolba, NASA ozone chief Robert Watson, and EPA administrator William Reilly called a press conference to announce the discovery that stratospheric ozone depletion is not confined to just Antarctica. Rather, we were told, it has spread to the United States where it will cause crop damage and dramatically accelerate incidents of skin cancer.

The actual numbers, still subject to vigorous debate in the scientific community (see Figure 11.1 for a reliable ozone record that shows a recent rise), indicated a midlatitude loss of 2 to 3 percent of stratospheric ozone. The associated increase in surface ultraviolet radiation is roughly equivalent to moving 30 miles south.

That figure immediately makes any argument about massive crop damage suspect. In fact, across the United States the natural flux of ultraviolet-B (UV-B) "counts" varies by a factor of 2, or 50 percent.

Figure 11.1
NOAA's Ozone Record from Mauna Loa

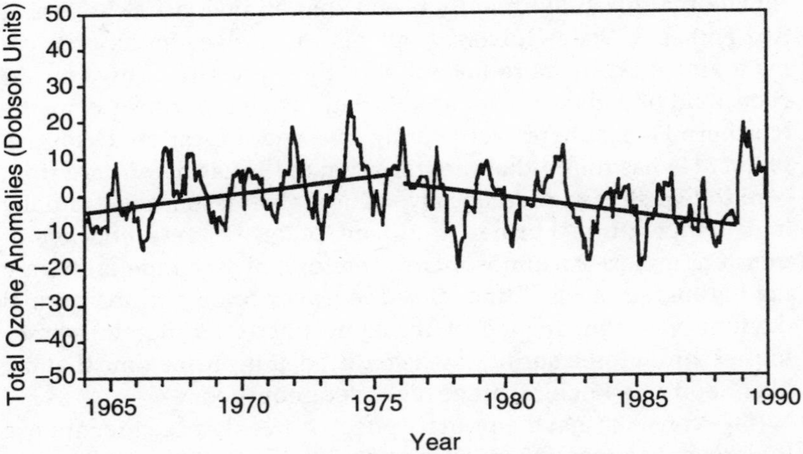

Note: Values for 1990 approached the maximum in the record.

Yet corn yields in the southern part of its range (Georgia and Florida) generally exceed those in the northernmost part (central Minnesota). Ditto for soybeans. And the state with the highest average corn yield is often Utah, whose high elevation basks in UV-B and whose population exhibits a high rate of basal cell skin cancer.

Further, despite the putative ozone decline, UV-B measurements have actually declined across the United States since they were first taken in 1974. If UV-B is the agent that causes basal cell skin cancer, it is harder to get it now than it was before, after adjusting for age and lifestyle. And what is the cause of less UV-B radiation's getting down to the earth's surface? Increased cloudiness and atmospheric turbidity caused by natural and anthropogenerated products.

Something is stopping the putative death rays, and the ubiquitous increase in cloudiness is a good candidate. What is truly peculiar about the 1991 announcement is that the very same organization that sponsored it (the United Nations) had recently published Ann Henderson-Sellers's (1989) cloudiness records. Those records make it quite clear that there have been dramatic increases in cloudiness that should more than compensate for a touted 3 percent decline in stratospheric ozone. The evidence is both direct—from cloud studies themselves—and indirect—from the day/night temperature distributions discussed in this book.

The real clincher is that measurements of high-elevation UV-B at mountain locations show slight increases, while the decreases are at low elevation. It should seem pretty obvious that something is interfering with the downward transmission of that radiation into the bottom 10,000 feet of the atmosphere.

Nonetheless, everyone agrees that stratospheric ozone during the late winter and early spring over Antarctica has declined precipitously since 1982. Although chlorofluorocarbons (CFCs) are implicated, the current theory (scientists prefer "hypothesis" or "model") still has to explain how the ozone depletion appeared so suddenly when CFCs had been increasing rapidly for decades. To do that, scientists have hypothesized that the Antarctic stratosphere has cooled so dramatically and suddenly (no reason given, although the stratosphere should cool some with an enhanced greenhouse effect) that a peculiar type of ice-crystal cloud forms in winter and freezes out stratospheric nitrogen, which would normally prevent a rapid ozone depletion. We should note that a natural ozone depletion also takes place during the polar night, because the formation of ozone occurs in the presence of ultraviolet radiation, or sunlight. Antarctica's isolation also helps, because it is surrounded by thousands of miles of water with no land or mountain ranges to disturb the ring of jet stream winds that howl circumferentially around the continent during the winter. The result is that very little air from the outside ever gets into Antarctic winter, so the ozone depletion is especially concentrated.

At any rate, even if we believe in the cloud accelerant, it is not cold enough away from the poles for that mechanism to work. Consequently, a new culprit has to be identified as the cause of depletion over, say, the United States. The UN pronouncement fingered sulfate aerosols.

There are two main sources of sulfate aerosols. The largest and most spectacular is an explosive volcano, such as Mt. Pinatubo in spring 1991. I wrote in a November 2, 1991, column for the *Washington Times* that an ozone scare would develop in early 1992 as the Pinatubo aerosol spread around the planet. But the UN/NASA/EPA data were collected long before that eruption. Therefore, the other mechanism for sulfate release had to be invoked: the burning of fossil fuel. That source of sulfate aerosol has been implicated as a cause of the cooling of the atmosphere and the benign expression of global warming.

Mankind is not about to stop burning fossil fuels, and the natural sources of sulfate aerosols have been present for billions of years. If the United Nations is going to implicate volcanic aerosol for ozone depletion, it had better admit that ozone holes are as common as big volcanoes. In fact, every time a good-sized volcano explodes when the stratosphere is cold (as it has been for most of the last 100 million years) and the coldest part of the stratosphere is over Antarctica (as it has been for the past 50 million years or so), a temporary, though dramatic, ozone hole should develop. If ozone holes are that common, the argument that they threaten the biosphere is absurd.

Now we have three competing environmental apocalypses: global warming, ozone depletion, and acid rain. The same compound (sulfate aerosol) that is associated with acid rain also fights global warming, increases cloudiness, blocks out UV-B, and is regularly pumped into the atmosphere by volcanoes. Sulfate aerosol may be the reason for a slight midlatitude stratospheric ozone depletion, but the associated increase in damaging UV-B radiation has not occurred where most things live, because of the screening effect of the compound and because of a concurrent increase in cloudiness that we would expect as a logical result of greenhouse enhancement. In addition, there is evidence that the same compound that causes global warming (CO_2) enhances plant growth at a rate that, in general, appears to far exceed any global effect of acid rain. Thus, the United Nations, by fingering sulfate aerosol as the cause of midlatitude ozone depletion, inadvertently hung the entire apocalypse machine on its own petard.

The interactions in the pollution system are quite obvious but are never factored into policy decisions. Consider the Clean Air Act, which is designed to remove sulfate aerosol from the atmosphere at a cost of $40 billion per year.

12. Newer Climate Models

> The question then is how well do the models do? The answer
> is they do pretty well. They do better than a factor of two.
> Now, if we made a major mistake in how we treat clouds
> and so forth, it would be very difficult to get it to come out
> as well as it does.
>
> —Stephen H. Schneider
> National Center for Atmospheric Research
> in Testimony to the U.S. House of
> Representatives, February 22, 1989

It is pretty clear that the GCMs of the mid-1980s were far from
complete. Those GCMs included oceans that did not mix vertically
or interact properly with the atmosphere, unrealistic cloud defini-
tions (a favorite: in an 8-by-10-degree latitude/longitude model,
when a thunderstorm develops—a feature that is usually a few
miles wide—it occupies an area the size of the entire Great Plains);
an earth that did not rotate with respect to the sun (and, therefore,
had no 24-hour day and night cycle; in those GCMs, it is always a
sunny day); and several other factors that would seem to render the
models inappropriate as underpinnings for economically dramatic
energy policies. Those models also suffered from the use of step-
wise (instantaneous) doubling of CO_2, rather than the low-order
exponential increase that occurs in the real world.

People who find fault with those GCMs are often accused of
setting up "straw men," because the models' limitations are so
well known within the scientific community. But the objections
are hardly "straw," because the models form the basis, and an
ostensibly scientific rationale, for the greatest experiment in the
central planning of energy in human history—a few trillion dollars
of expenditure, exacted mainly from a nation that routinely runs a
$250 billion annual deficit. As the IPCC's Chris Folland (1991) told
the Asheville scientists, "The data don't matter," and, "we're [the
United Nations is] basing recommendations upon the models." The

169

older models projected a mean equilibrium warming for a doubling of atmospheric CO_2 of 4.2°C (7.6°F) with maximum warming of as much as 18°C (32.4°F) during the north polar winter and less significant warming (approximately 2.0°C or 3.6°F) over tropical oceans. The illustration in Chapter 4, from the NCAR GCM, is typical.

The projections of those models are changing, and many are cooling. With a modified ice-water interaction between clouds, projected warming in a new United Kingdom Meteorological Office model (UKMO) dropped from 5.2°C (9.4°F) to 1.9°C (3.4°F), according to a paper published in *Nature* by J. F. B. Mitchell. When the oceans and the atmosphere were considered together, the net warming in the National Center for Atmospheric Research (NCAR) model dropped to 1.6°C (2.9°F) for 30 years after an instantaneous doubling of CO_2, compared to 3.7°C (6.7°F) in an earlier equilibrium calculation (see Figure 12.1). Although it is unclear whether the *eventual* temperatures produced by those two models are the same, it is obvious that when the oceans and the atmosphere interact, warming is dramatically delayed.

In a slightly different version (Washington and Meehl 1989) the NCAR model employs a greenhouse enhancement that mimics what has occurred in the past, say, 30 years, with a realistic 1 percent per year increase in equivalent CO_2 (see Figure 12.2). Washington and Meehl reported that after 30 years of increases of that magnitude, the global temperature should rise about 0.7°C (1.3°F). That rise could be considered analogous to, say, what occurred during the period between 1950 and 1980, and it is noteworthy that the net warming over *any* 30-year period since the major portion of greenhouse enhancement began is *less than one-half* of what was projected by this, the coolest of the major GCMs.

Figure 12.2 details the distribution of December–February temperature after 25 years of realistic greenhouse gas enhancement. Roughly speaking, the map could be analogous to a five-year aggregate in the 1980s. The most significant projected anomalies are the 2°C to 4°C (3.6°F to 7.2°F) warming of the northern half of North America and the 3°C to 6°C (5.4°F to 10.8°F) cooling of the North Atlantic. Neither occurred, although Atlantic temperatures dropped, as noted earlier. The projected North Atlantic cold anomaly does not appear in the next five-year average, which indicates

Figure 12.1

NEW NCAR GCM RESULTS FOR 30 YEARS AFTER DOUBLING OF CO_2

Surface Air Temperature Differences, (December–February)

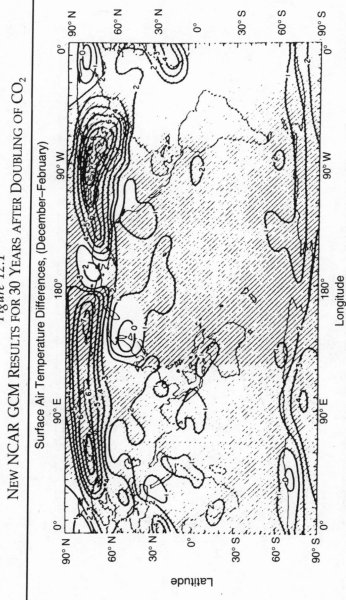

NOTE: This figure shows the "new" NCAR model's results for coupled atmosphere-ocean readings for 30 years after a "shock doubling" of CO_2. Compare this figure to the NCAR model in Chapter 4.

171

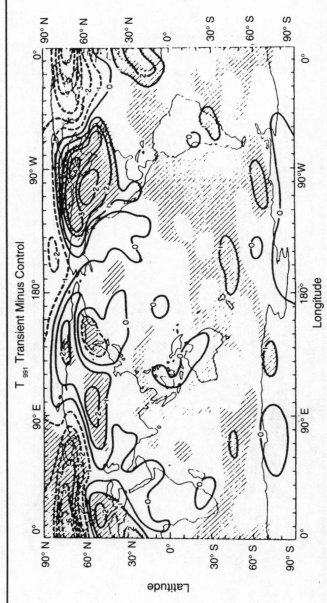

Figure 12.2
New NCAR GCM Results for Years 26–30

Note: This figure shows a five-year average of expected changes from the background climate for years 26–30 of an NCAR model in which the greenhouse enhancement is gradually increased in a fashion that is more realistic than the "shock doublings." Such a period might be analogous to December–February temperatures for five years in the 1980s. The projected large anomalies were not observed in reality.

that the model is producing a lot of "noise," or spurious random variability. In fact, on five-year scales, coolings of the magnitude projected over such large areas are virtually nonexistent in the historical record of the past 100 years.

The Princeton group recently published two new versions of their model (Manabe et al. 1991). One reports a net equilibrium warming for a doubling of CO_2 with a coupled atmosphere-ocean model of 1.8°C (3.2°F), compared to 4.3°C (7.7°F) in the more primitive calculation. The net warming projected for the Southern Hemisphere is approximately 1.0°C (1.8°F), and 2.5°C (4.5°F) warming is projected for the Northern Hemisphere. The observed behavior of the past 50 years—in which the Southern Hemisphere shows a more greenhouse-like warming than the Northern Hemisphere—seems inconsistent with this model's projection unless there has been some mitigation of Northern Hemisphere warming by means not included in the modeling process. In the 1991 article in the *Journal of Climate*, the overall equilibrium warming for a doubling of CO_2 is back up to 4.0°C (7.2°F), but the time scale has been lengthened. The warming is probably stretched out in time because this generation of the model uses a much more realistic version of ocean dynamics and their coupling to the atmosphere than did earlier ones. It also uses an increase in greenhouse enhancement that is much more realistic than instantaneous doubling. It is noteworthy that all of the warming of more than 4.0°C (7.2°F) on this annual map is confined to very high northern latitudes. That means that more than half of the significant warming is projected for winter, when the sun is below the horizon in polar night. Figure 12.3 shows a map of that model's forecast for the year 2040, as well as the NCAR map for 30 years after a (unrealistic) shock doubling of CO_2.

The *Journal of Climate* paper also reports a "free running" model in which the greenhouse is not enhanced. Figure 12.4 shows behavior in the Northern Hemisphere superimposed on a rise of 1.4°C (2.5°F) in the past 100 years, which is one-half of the 2.8°C (5.0°F) warming that the model calculates will occur by the time CO_2 doubles (we are halfway there). If we "anchor" the two plots together at the beginning of the temperature history in 1890, beginning around 1940 a considerable disparity emerges between the predicted and the observed value. There seems to be no way around

Figure 12.3
TEMPERATURE CHANGES IN 2040 PREDICTED BY GFDL
MODEL

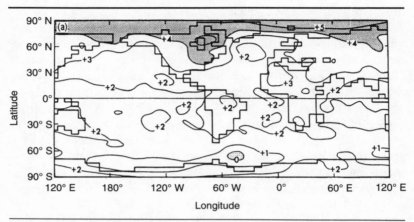

NOTE: The new Princeton (GFDL) model, published in the *Journal of Climate* in 1991, shows changes from the background that might be expected around 2040.

Figure 12.4
OBSERVED VS. CALCULATED CHANGES IN TEMPERATURE OF
NORTHERN HEMISPHERE

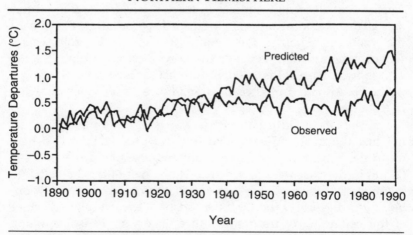

NOTE: This figure illustrates the inability of the best GCMs (in this case, the GFDL model) to simulate the last 100 years.

that error. If we force the predicted and observed temperatures to be the same now, the model will make the early part of this century far too cold.

In the new models, less warming is calculated when either the oceans or the clouds are simulated in a more realistic fashion. Therefore, an improved model that incorporated both clouds and oceans would produce even less net warming.

On February 22, 1989, before the new results were published, NCAR scientist Steven Schneider (who is, according to newspaper surveys, the most quoted authority on global warming) testified before the House Subcommittee on Energy and Power: "The question then is how well do the models do? The answer is they do pretty well."

On September 14, 1989, after the new, cooler UKMO model came out, he told the *Washington Post*, "It is going to be another decade or two before we have answers that are credible." Thus, it seems the warmer models "do pretty well," and the cooler ones do not provide "answers that are credible."

Perhaps we were too easy on the Intergovernmental Panel on Climate Change or on our elected officials when we noted that statements to the effect that "all scientists agree the greenhouse effect is real," or that "the greenhouse effect is an accepted scientific fact" are about as profound a revelation as the fact that the earth is round. One peculiar fact that appears in nearly every scientific publication by GCM groups is that results are presented in a fashion that indicates that, indeed, the earth is flat.

In almost every publication of GCM results that I have seen in the refereed literature, in every UN report detailing results, and in every presentation to the U.S. Congress that I have witnessed, the amount of significant projected warming (say, more than 4°C or 7.2°F) has been dramatically distorted by the use of map projections that assume the area in each band of latitude is equal. Of course, in reality, the area covered by 10 degrees of latitude is much greater near the equator than it is at higher latitudes. In fact, the total area in the highest latitude bands—80°–90° N—is very small: less than 1 percent of the planet. North of 70° it is about 3 percent.

Consider the new NCAR models's December–February (Northern Hemisphere winter/Southern Hemisphere summer) projection

(Figure 12.1) for 30 years after a shock doubling of CO_2. The areas of warming of more than 4°C (7.2°F) still appear to be very large, because they are in the high latitudes. (The analogous Northern Hemisphere summer/Southern Hemisphere winter model shows no areas of warming of 4.0°C or 7.2°F in either hemisphere.) But that projection assumes the area from 70° to 90° N, which shows almost all of the warming of more than 4.0°C (7.2°F), is equal to the area in any other 20° latitude band. It assumes that band contains two-ninths (22 percent) of our hemisphere, when in fact it contains only about 6 percent. The distortion factor in high-latitude warming is therefore on the order of 300 percent.

In fact, when measured on an actual globe, the area of projected warming of more than 4.0°C (7.2°F) in this model—on an annual basis (after adjusting for the fact that there are no warmings of that magnitude projected for Northern Hemisphere summer)—is less than 5 percent. However, it is doubtful that any nonprofessional observer of those results—including our elected officials—could possibly realize the amount of distortion that is inherent when the earth is flattened.

In the new coupled ocean-atmosphere GFDL model using realistic trace gas inputs, the temperature projection for circa 2040 suffers from even more of the same. Most of the warming of more than 4.0°C (7.2°F) is confined to latitudes north of 75°, or about 2 percent of the planet. It surely looks like a larger area in the *Journal of Climate* (see Figure 12.3).

Figure 12.5 shows how the areas of projected warming of more than 4°C (7.2°F) in that model would look to a spaceship parked over the North Pole. It is quite apparent that large "global" warming looks a lot more "local" from that vantage point.

I have done the same for the NCAR winter run for 30 years after a shock doubling of CO_2. Figure 12.6 shows the areas of four or more degrees of warming from north and south polar projections and from two equatorial projections, at 90° W (North and South America) and 90° E (Asia). In three of them virtually no warming of the required magnitude is visible. Most of the warming is reserved for the north polar projection.

The one lower latitude area that does show a patch of warming of more than 4°C (7.2°F) is the Sahara Desert. Along with the poles, the Sahara is one of the most greenhouse gas–impoverished regions

Figure 12.5
NORTH POLAR PROJECTION FOR NEW GFDL MODEL

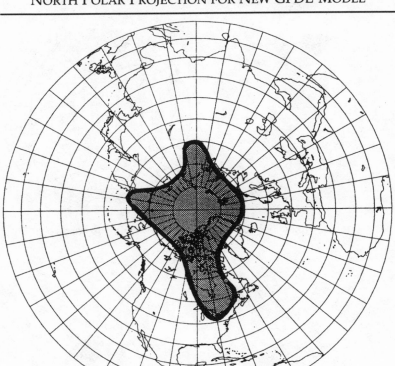

NOTE: This figure shows a north polar projection of areas that will experience more than 4°C (7.2°F) of warming by 2040 in the new GFDL model that has more realistic oceans and greenhouse enhancements. From this point of view, global warming becomes pretty local and is restricted to an environment that is profoundly cold.

of the planet because of the profound lack of water, which is the big greenhouse enhancer. Adding a small increment of greenhouse gas to such a greenhouse-poor environment will surely result in a rise in temperature, but few will be there to notice. In the overpopulated and ecologically stressed Sahel (immediately to the south), aridity will probably increase, too, but I doubt that things could get

Figure 12.6
FOUR NCAR MODEL PROJECTIONS FOR WINTER

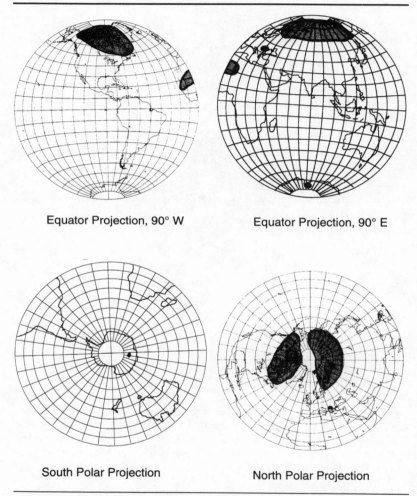

Equator Projection, 90° W

Equator Projection, 90° E

South Polar Projection

North Polar Projection

NOTE: These four maps show areas of warming of 4°C (7.2°F) or more projected for 30 years after a shock doubling of CO_2 in the NCAR 1989 model.

much worse there than they are now after two decades of extreme drought.

While the climate models are probably still too warm in general, almost all of the warming in the newer ones is confined to high-latitude winter. Because the sun is either at or beneath the horizon, almost all of the warming of more than 4°C (7.2°F) is projected for evening or night.[1]

Thus, even though the models do not include the anthropogenerated particulates, they show that almost all of the strong warming of the next century will be in the winter, at night, or in the highest latitudes (or some combination of those three). Adding the particulate effect is likely to push warming further in those directions.

A look back to Arrhenius' 1896 paper (see Chapter 1) is instructive. Given the history of the past 50 years, it is pretty apparent that his only mistake was an overestimation of total enhanced greenhouse warming. The rest—that the warming would tend to be reflected as a decline in the daily and seasonal temperature ranges, that it would be greater at high latitudes and in winter, and that it would, therefore, concentrate at night—is as correct today as it was 100 years ago and as inconsequential to the fate of the planet.

1. Here is an exercise that is quite convincing. Purchase a globe and paint on it all areas of 4.0°C (7.2°F) or more of warming from the NCAR winter model shown in Figure 12.1 (there are none in the summer version). Now place the globe near a light bulb. Tilt the globe so that the Tropic of Capricorn (23.5° S latitude) is lined up with the sun. That is the tilt of the planet's axis with respect to the sun in winter. The sun shines overhead at the Tropic of Capricorn on December 21, the first day of winter. Now spin the globe. How much of the area of 4.0°C (7.2°F) or more of warming does the sun "see," and how much remains in the shadow of high northern latitude polar night? The answer is that almost all warming demonstrably occurs at night.

13. Consensus?

One of the general tenets of those who subscribe to the Popular Vision is that there is a consensus among scientists that the end is at hand. That view is understandable, inasmuch as there has been produced only one television documentary that takes the opposite viewpoint (the previously described "Greenhouse Conspiracy"), whereas there have been dozens of scare stories, such as "Race to Save the Planet" (starring Meryl Streep), James Burke's "After the Warming" (complete with flat earth maps), and NOVA's "Hot Enough for You?" (ditto), all of which were showcased by the Public Broadcasting Service. PBS refused to broadcast "The Greenhouse Conspiracy" nationally because it was "too biased."

Three Surveys

Three recent surveys document that the true consensus among scientists does not coincide with the Popular Vision.

The Science and Environmental Policy Project

Is the consensus of scientists roughly allied with the consensus of most television documentarians? S. Fred Singer, who is with the Science and Environmental Policy Project, recently surveyed three representative groups: contributors to the IPCC report, reviewers of that report, and a third group variously referred to as the Phoenix group or the Gang of 25 that consists of 24 scientists who met in Phoenix, Arizona, in 1990 to draft a research program designed to test the hypothesis that climate apocalypse is not at hand. Although the résumés and track records of the members of the latter group cannot be distinguished from those of either the contributors to or the reviewers of the IPCC report, there was very little overlap between the Phoenix and IPCC groups. It is hard to believe that resulted from random selection processes, although such claims have been made.

The respondents to Singer's October 1991 survey, which was published in the *Wall Street Journal*, came almost equally from the three groups. In a subsequent article, Singer noted that the IPCC reviewers tended to hold views closer to those of the Phoenix group than did the contributing authors.

The key to the IPCC document is its Policymakers Summary, written by a small steering group, that emphasizes "certainty" about the greenhouse effect (again, "all scientists agree"). Forty percent of the IPCC reviewers and authors combined believed that summary conveyed a misleading impression of the nature of upcoming climate change. Almost all of the Phoenix group concurred.

All three groups strongly agreed with the proposition that the GCMs could not reproduce the climate behavior of the past century. Further, Singer wrote that "60 percent of the IPCC group, and all of the Phoenix group, believed that current global circulation models do not accurately portray the real atmosphere-ocean system. Yet these models form the only basis for predicting a catastrophic warming in the next century."

The Gallup Poll

On February 13, 1992, the Gallup organization released a poll of members of the American Geophysical Union and the American Meteorological Society. Members of those two professional societies are most likely to be actively involved in research on climatic change. Only 17 percent of the respondents thought that the warming of the 20th century, popularly thought to be consistent with an enhanced greenhouse, was the result of a change in the greenhouse effect.

Greenpeace

In a Greenpeace survey of IPCC scientists and researchers who had published on issues relevant to climate change in *Science* or *Nature* during 1991, only 13 percent of the respondents said that continuing the same energy mix would result in a runaway greenhouse effect; 32 percent said that might possibly be the case; and 47 percent believed that business as usual would "probably not" induce a runaway greenhouse effect. Greenpeace had a unique interpretation of the results of its survey; it called the majority of responses an "as-yet poorly expressed fear." Several IPCC scientists, most of whom are not "apocalyptics," have indicated that they were not contacted by Greenpeace.

The total number of scientists who answered Singer's survey was around 40 (not all answered all questions), which might seem like a pretty small, and perhaps unrepresentative, sample. In fact, 40 scientists is a pretty large slice of the active research community that works either with climate histories or climate models (a few do both). In 1988 the NOAA attempted to develop a collaborative grouping, called the Climate Trends Panel, of all such scientists in the United States (there were also one or two from outside the country). The total membership numbered around 60. Calls for papers to the American Meteorological Society's Applied Climatology section, which meets every other year, usually bring in about 60 manuscripts, and several authors submit more than one.

So where do we find the "hundreds of scientists" who are usually cited as representing this or that consensus that the time to act is now. They are not members of the climate science community; there simply are not that many of us. Those who would build an "action-now" consensus have recourse to a related resource: scientists who are not climatologists at all but are in distantly related fields. While they might not be in a good position to pass judgment on GCMs, those scientists know that GCM-driven warming scenarios are the vehicle for funding research that heretofore may not have been supported.

A Letter

In February 1991, more than 50 scientists, most of whom are or were in leadership positions in the American Meteorological Society, signed a letter containing the following language.

> As independent scientists, researching atmospheric and climate problems, we are concerned by the agenda for the United Nations Conference on Environment and Development (to be held in June 1992 at Rio de Janeiro) being developed by environmental activist groups and certain political leaders. . . .
> [The] policy initiatives derive from highly uncertain scientific theories. They are based on the unsupported assumption that catastrophic global warming follows from the burning of fossil fuel and requires immediate action. We do not agree. . . .
> We are disturbed that activists, anxious to stop energy and economic growth, are pushing ahead with drastic policies

without taking note of recent changes in the underlying science. We fear that the rush to impose global regulations will have catastrophic impacts on the world economy, standard of living, and health care, with the most severe consequences falling upon developing countries and the poor.

The Phoenix Group

On October 12–13, 1990, the Laboratory of Climatology at Arizona State University sponsored a meeting of scientists to define a research agenda that would complement other ongoing studies of climatic change, but with a twist. The working hypothesis of the group was to be that we may not be headed toward the apocalyptic changes feared by many.

Twenty-eight scientists of national and international reputation were invited, and only four declined the invitation—an unusually high acceptance rate for such a conference. There is clearly a strong professional motivation to explore the emerging view of neutral or possibly beneficial climatic change.

The results of that meeting are detailed in "Global Climatic Change: A New Vision for the 1990s," available from Dr. Robert Balling of Arizona State University. Most interesting is the apparent consensus that is detailed in the report.

> The consensus of the scientists in this research prospectus is that there is considerable evidence that the impact of future climatic change may be neutral or even beneficial. The lines of evidence include
>
> - *The Magnitude of Observed Warming.* The historical record of observed temperature change suggests that global warming for a doubling of carbon dioxide will be far below the 4.2° that fuels the Popular Vision.
> - *The Timing of Observed and Projected Warming.* More refined climate models tend to project most of their warming to occur in high latitude winter, which partitions most warming into the night. This prevents most of the deleterious effects and in fact lengthens growing seasons. The warmth of 1990 was consistent with this projection as were the world-averaged record crop yields.
> - *The Growth Enhancement Caused by Carbon Dioxide.* Carbon dioxide is currently a limiting nutrient for plants, and a voluminous scientific literature demonstrates enhanced

growth and water use efficiency as its concentration increases. In fact, except for the height of the ice ages, both global temperature and CO_2 concentration are currently near their lowest values for the last 100 million years.

Later, the report states that "this document details a series of research proposals built around a central hypothesis: *the Popular Vision is wrong*" (italics in original).

In response to a request from one of the members of the group, all conference participants were polled as to their level of agreement with the three bulleted statements just quoted. Participants were asked to rate their strongest agreement at +10 and their strongest disagreement at −10.

The mean score was +8. Only three of the participants were not reached for response, and only three of the respondents gave agreement scores of less than zero. The respondents all held doctoral degrees and were employed, either as faculty or as professional staff, by the following institutions: Woods Hole Oceanographic Institute, Arizona State University, the University of Virginia, the Smithsonian Institution, Lawrence Livermore National Laboratory, the U.S. Department of Commerce, Columbia University, M.I.T., Colorado State University, the U.S. Department of Agriculture, Yale, and Rutgers.

Given the results of the surveys discussed in this chapter and the track records of the individuals involved, it is simply impossible to find any scientific consensus supporting the Popular Vision of climate disaster. Rather, the consensus is the opposite: the Popular Vision is unscientific.

14. The Long-Range Forecast

Making a forecast is easy. Being right is the hard part.
—Reid Bryson
University of Wisconsin

Bryson's remonstrance to his climatology class notwithstanding, no book dealing with climate change is complete without a forecast of what will happen in the coming decades.

The Temperature Record Will Warm

Warming must occur because the number of people surrounding weather stations will continue increase. Whether or not the globe truly warms, each succeeding year will tend to show some warming attributable to increased population. Thus, we can, at least in the near term, expect to see news stories every January about how the preceding year was the warmest, or nearly the warmest, year in the past century. The annual climate record that comes over the teletype first and receives the first analysis is usually from cities and therefore reports temperatures that are too warm.

The Planet Will Continue to Warm

There is little doubt that we have had some warming over the past 100 years—probably between one-quarter and one-half of a degree (C). In the absence of good information, prognosticators know that the way to come up with a forecast that is more accurate than the one flipping a coin would produce is to say that the future will continue to be like the past. Further, the greenhouse enhancement is likely to contribute to warming, too, even though there are competing inhibitors, such as sulfate aerosol and other cloud enhancers. The warming that will have occurred between 1900 and the time CO_2 effectively doubles in the next century will be on the order of 1.0°C to 1.5°C (1.8°F to 2.7°F); in other words, temperatures will be a little less than one degree warmer than they are today.

Illusion of Rapid Warming

The combination of El Niño events, along with artificial urban warming, and a slow real warming will create the illusion of some brief periods of apparently rapid warming. We saw what happened from 1917 through 1921, and there is no reason it cannot happen again. A good time for a repetition might be in the mid-1990s, after Mt. Pinatubo's ash disperses, and a strong El Niño roars in.

Global Warming Can Elect a President

Suppose the following: Pinatubo's ash cloud lingers through 1993, and a big El Niño cranks up in 1994. The California drought reappears, and drought also shows up in Texas. Politicians who have established a track record on global warming, such as Al Gore or George Mitchell, hit the stump: Ronald Reagan caused the problems by not supporting solar energy and windmills. George Bush caused the drought by inaction in the face of the international consensus on global warming. All scientists agree that the greenhouse effect is real. Dan Quayle said the solution to global warming was to burn natural gas, which emits carbon dioxide, and he's running for president. California and Texas are always close in a presidential election, and the winner of both is halfway to the White House.

Emissions Reduction Treaty

The United States will agree to a carbon-emissions reduction treaty. The United States did not sign that agreement at the Rio de Janeiro meeting in June 1992, but it awaits only the inevitable brief period of rapid warming. Further, the environmental lobby—the largest lobby in our nation's history—will not be satisfied until the United States agrees to such a treaty. The cost will be several trillion dollars to a nation already deeply in debt.

20 Years from Now

Twenty years from now, we will more fully appreciate the effects of modest warming and CO_2 enhancement. We will see longer growing seasons, summer temperatures that do not change much (except that summers will tend to be longer), warmer nights, not much change in day temperatures, and a greener planet.

188

Remorse

The costs of a carbon treaty will prove to be enormous, but the revenue generated will be addictive to government. Large taxes, once legislated, will not go away quietly into the (warmer) night.

On warm nights we will look out on green fields and wonder how we could have been so foolish. Historians in the 21st century will note that, even by the mid-1980s, the best available data indicated that the then Popular Vision of climate catastrophe was a failure.

References

Angell, J. K., 1990: Variation in global tropospheric temperature after adjustment for the El Niño influence, 1958–89. *Geophys. Res. Let.* **17**, 1093–96.

Angell, J. K., 1990: Variation in United States cloudiness and sunshine duration between 1950 and the drought year of 1988. *J. Climate* **3**, 296–306.

Balling, R. C. (distributor), 1990: *Global Climatic Change: A New Vision for the 1990s.* Proceedings of a Research Symposium, Laboratory of Climatology, Arizona State University. 26 pp.

Balling, R. C., and S. B. Idso, 1989: Historical temperature trends in the United States and the effect of urban population growth. *J. Geophys. Res.* **94**, 3359–63.

Balling, R. C., and S. B. Idso, 1990: 100 years of global warming? *Envi. Consv.* **17**, 165.

Balling, R. C., J. A. Skindlov, and B. H. Phillips, 1990: The influence of increasing summer mean temperatures on extreme maximum and minimum temperatures in Phoenix, Arizona. *J. Climate* **3**, 1491–94.

Bottomley, M., C. K. Folland, J. Hsuing, R. E. Newell, and D. E. Parker, 1990: *Global Ocean Surface Temperature Atlas.* U.K. Met. Office, Bracknell, UK. 20 pp. +313 Plates.

Bradley, R. S., H. F. Diaz, J. K. Eisheid, P. D. Jones, P. M. Kelly, and C. M. Goodess, 1987: Precipitation fluctuations over the Northern Hemisphere land areas since the mid 19th Century. *Science* **237**, 171–75.

Bruhl, C., and P. J. Crutzen, 1989: On the disproportionate role of tropospheric ozone as a filter against solar UV-B radiation. *Geophys. Res. Let.* **16**, 703–9.

Bryson, R. A., and G. J. Dittberner, 1976: A non-equilibrium model of hemispheric mean temperature. *J. Atm. Sci.* **33**, 2094–2106.

Cess, R. E., 1989: presentation to Department of Energy Research Agenda Workshop 4/25/89, Germantown, Md.

Charlson, R. J., S. E. Schwartz, J. M. Hales, R. D. Cess, J. A. Coakley, J. E. Hansen, and D. J. Hoffmann, 1992: Climate forcing by anthropogenic aerosols. *Science* **225**, 423–30.

Craig, H., C. C. Chou, J. A. Wehlan, C. M. Stevens, and A. Engelmeier, 1988: The isotopic composition of methane in polar ice cores. *Science* **242**, 1535–39.

Cullis, C. F., and M. M. Hirschler, 1980: Atmospheric sulfur: Natural and man-made sources. *Atm. Envi.* **14**, 1263–78.

191

REFERENCES

Department of Energy, 1990: *Trends '90,* Carbon Dioxide Information Analysis Center, Oak Ridge, Tenn. 257 pp. The temperature record used in this paper is by Jones et al. and is detailed on page 194 of the publication.

Domack, E. W., A. J. Jull, and S. Nakao, 1991: Advance of East Antarctic outlet glaciers during the Hypsithermal: Implications for the volume state of the Antarctic ice sheet under global warming. *Geology* **19,** 1059–62.

Ellsaesser, H. W., 1990: *Planet Earth: Are Scientists Undertakers or Caretakers?* Lawrence Livermore National Laboratory. 20 pp.

Ellsaesser, H. W., M. C. MacCracken, J. J. Walton, and S. L. Grotch, 1986: Global climatic trends as revealed by the recorded data. *Rev. Geophys.* **24,** 745–92.

Elms, J., and R. Quayle, 1992: Multi-decade sea surface temperature trends in American waters afflicted by coral bleaching. American Meteorological Society, *Third Symposium on Global Change Studies,* Atlanta, Ga., pp. 98–101.

Grotch, S. L., 1991: A statistical intercomparison of temperature and precipitation predicted by four general circulation models. PP 3-16 *In* Schlesinger, M. E. (ed.), *Greenhouse-Gas-Induced Climatic Change: A Critical Appraisal of Simulations and Observations.* Elsevier, N.Y., 615 pp.

Hansen, J. E., 1988: Testimony to the U. S. Senate, Committee on Energy and Natural Resources, June 23, 1988.

Hansen, J., I. Fung, A. Lacis, D. Rind, S. Lebedeff, R. Ruedy, G. Russell, and P. Stone, 1988: Global climate changes forecast by the Goddard Institute for Space Studies three-dimensional model. *J. Geophys. Res.* **93,** 9341–64.

Hansen, J. E., and A. A. Lacis, 1990: Sun and dust versus greenhouse gases: An assessment of their relative roles in global climate change. *Nature* **346,** 713–718.

Hansen, J. E., and S. Lebedeff, 1988: Global surface air temperatures: Update through 1987. *Geophys. Res. Let.* **15,** 323–26.

Hansen, J. E., A. Lacis, D. Rind, G. Russell, P. Stone, I. Fung, R. Ruedy, and J. Lerner, 1984: Climate sensitivity: Analysis of feedback mechanisms. *Geophys. Mono. Ser.* **29,** 130–63.

Hanson, K., G. A. Maul, and T. R. Karl, 1989: Are atmospheric "greenhouse" effects apparent in the climate record of the contiguous United States? *Geophys. Res. Let.* **16,** 49–52.

Henderson-Sellers, A., 1986: Increasing cloud in a warming world. *Climatic Change* **9,** 267–309.

———, 1989: North American total cloud amount variations this century. *Global Planet. Change* **1,** 175–94.

Intergovernmental Panel on Climate Change (IPCC), 1990: *Policymakers Summary of the Scientific Assessment of Climate Change.* World Meteorological Organization, United Nations Environment Programme. 200 pp.

192

———, 1992. *Climate Change 1992. The Supplementary Report to the IPCC Scientific Assessment.* World Meteorological Organization, United Nations Environment Programme. 39 pp.

Idso, S. B., 1989: *Carbon Dioxide and Global Change: Earth in Transition,* IBR Press, 292 pp.

Idso, S. B., 1990: Evidence in support of Gaian climate control: Hemispheric temperature trends of the past century. *Theor. Appl. Clim.* 42, 135–37.

Idso, S. B., and R. C. Balling, Jr., 1991: Surface air temperature response to increasing global industrial productivity: A beneficial greenhouse effect? *Theor. Appl. Climatol.* **44,** 37–41.

Idso, S. B., R. C. Balling, and R. S. Cerveny, 1990: Carbon dioxide and hurricanes: Implications of Northern Hemisphere warming for Atlantic/Caribbean storms. *Met. and Atm. Phys.* **42,** 259–63.

Jones, P. A., 1991: Historical records of cloud cover and climate for Australia. *Aust. Meteor. Mag.* **39,** 181–89.

Jones, P. D., P. Groisman, M. Coughlan, N. Plummer, W. -C. Wang, and T. R. Karl, 1990: How large is the urbanization bias in large area-averaged surface air temperature trends? *Nature* **347,** 169–72.

Kalkstein, L. S., P. C. Dunne, and R. S. Vose, 1990: Detection of climatic change in the western North American Arctic using a synoptic climatological approach. *J. Climate* **3,** 1154–67.

Karl, T. R., and P. D. Jones, 1989: Urban bias in area-averaged surface air temperature trends. *Bull. Amer. Met. Soc.* **70,** 265–70.

Karl, T. R., H. F. Diaz, and J. Kukla, 1988: Urbanization: Its detection and effect in the United States climate record. *J. Climate* **1,** 1099–1123.

Karl, T. R., R. G. Baldwin, and M. G. Burgin, 1988: *Historical Climatology Series* **4–5,** National Climatic Data Center, Asheville, N.C., 107 pp.

Karl, T. R., G. Kukla, V. N. Razuvayev, M. G. Changery, R. G. Quayle, R. R. Heim, Jr., D. R. Easterling, and C. B. Fu, 1991: Global warming: Evidence for assymetric diurnal temperature change. *Geophys. Res. Let.* **18,** 2252–56.

Kaufmann, Y. J., R. S. Frazer, and R. L. Mahoney, 1991: Fossil fuel and biomass burning effect on climate-heating or cooling? *J. Climate* **4,** 578–88.

Kauppi, P. E., K. Mielikainen, and K. Kuusela, 1992: Biomass and carbon budget of European forests, 1971 to 1990. *Science* **256,** 70–74.

Kiehl, J. T., and R. E. Dickinson, 1987: A study of the radiative effects of enhanced atmospheric CO_2 and CH_4 on early earth surface temperatures. *J. Geophys. Res.* **92,** 2991–98.

LaMarche, V. C., D. A. Greybill, H. C. Fritts, and M. R. Rose, 1984: Increasing atmospheric carbon dioxide: Tree ring evidence for growth enhancement in natural vegetation. *Science* **223,** 1019–21.

Lorenz, E. N., 1984: Irregularity: A fundamental property of the atmosphere. *Tellus* **36A,** 98–110.

Lorius, C., J. Jouzel, C. Ritz, L. Merlivat, N. I. Barkov, Y. S. Korotkevich, and V. M. Kotlyakov, 1987: A 150,000 climatic record from Antarctic ice. *Nature* **329**, 591–96.

MacCracken, M. C., and G. J. Kukla, 1985: *Detecting the Climatic Effects of Increasing Carbon Dioxide*, U.S. Department of Energy DOE/ER-1235, pp. 163–76.

Manabe, S., and R. T. Wetherald, 1980: On the distribution of climate change resulting from an increase in the CO_2 content of the atmosphere. *J. Atmos. Sci.* **37**, 99–118.

Manabe, S., R. J. Stouffer, M. J. Spelman, and K. Bryan, 1991: Transient responses of a coupled ocean-atmosphere model to gradual changes of atmospheric CO_2: Part 1: Annual mean response. *J. Climate* **4**, 785–818.

Mayeux, H. S., and H. Johnson, 1986: *Causes and Consequences of Vegetation Changes on Rangeland*, USDA/ARS Research Project Statement, CRIS Work Unit #6206-20110-004.

Mayewski, P. A., W. B. Lyons, M. J. Spencer, M. S. Twickler, C. F. Bock, and S. Whitlow, 1990: An ice-core record of atmospheric response to anthropogenic sulphate and nitrate. *Nature* **346**, 554–56.

Michaels, P. J., 1982: The response of the "Green Revolution" to climatic variability. *Cli. Change* **4**, 255–71.

Michaels, P. J., D. E. Sappington, D. E. Stooksbury, and B. P. Hayden, 1990: Regional 500mb heights and U.S. 1000–500mb thickness prior to the radiosonde era. *Theor. Appl. Clim.* **42**, 149–54.

Miller, G. H., and A. de Vernal, 1992: Will greenhouse warming lead to Northern Hemisphere ice-sheet growth? *Nature* **355**.

Mitchell, J. F. B., 1983: The seasonal response of a general circulation model to changes in CO_2 and sea temperature. *Quart. Jour. Royal Meteor. Soc.* **109**, 113–53.

Mitchell, J. F. B. , C. A. Senior, and W. H. Ingram, 1989: CO_2 and climate: A missing feedback. *Nature* **341**, 132–34.

Neftel, A., E. Moor, H. Oeschger, and B. Stauffer, 1985: Evidence from polar ice cores for the increase in atmospheric CO_2 in the past two centuries. *Nature* **315**, 45–47.

Oort, A. H., Y. H. Pan, R. W. Reynolds, and C. Ropewlski, 1989: Historical trends in surface temperature over the oceans based on the COADS. *Climate Dynamics* **2**, 29.

Ramanathan, V., R. D. Cess, E. F. Harrison, P. Minnis, G. B. R. Barkstrom, E. Ahmad, and D. Hartmann, 1989: Cloud-radiative forcing and climate: Results from the Earth Radiation Budget Experiment. *Science* **243**, 53–67.

Ramaswamy, V., M. D. Schwarzkopf, and K. P. Shine, 1992: Radiative forcing of climate from halocarbon-induced global stratospheric ozone loss. *Nature* **355**, 810–12.

Raynaud, D., and J. M. Barnola, 1985: An Antarctic ice core reveals atmospheric CO_2 variations over the past few centuries. *Nature* **315**, 309–11.

Rind, D., R. Goldberg, J. Hansen, C. Rosensweig, and R. Ruedy, 1990: Potential evapotranspiration and the likelihood of future drought. *J. Geophys. Res.* **95,** 9983–10004.

Robock, A., 1983: *Second Conference on Climate Variations,* American Meteorological Society, New Orleans, La.

Rogers, J. C., 1989: *Proc. 13th Annual Climate Diagnostics Workshop, NOAA;* Available from NTIS.

Sansom, J., 1989: Antarctic surface temperature time series. *J. Climate* **2,** 1164–72.

Schlesinger, M. E., 1984: Climate model simulation of CO_2 induced climatic change. *Adv. Geophys.* **26,** 141–235.

Schneider, S. H., and R. Chen, 1980: Carbon dioxide warming and coastline flooding: Physical factors and climate impact. *Ann. Rev. Energy* **5,** 107–35.

Scotto, J., G. Cotton, F. Urbach, D. Berger, and F. Fears, 1988: Biologically effective ultraviolet radiation: Surface measurements in the United States. *Science* **239,** 762–63.

Seaver, W. L., and J. E. Lee, 1987: A statistical examination of sky cover changes in the contiguous United States. *J. Clin. App. Meteor.* **26,** 88–95.

Singer, S. F., C. Starr, and R. Revelle, 1991: What to do about greenhouse warming: Look before you leap. *Cosmos* **1** (1).

Sionit, N., H. Hellmers, and B. R. Strain, 1980: Growth and yield of wheat under carbon dioxide enrichment and water stress conditions. *Crop. Sci.* **20,** 687–90.

Spencer, R. W., and J. R. Christy, 1990: Precise monitoring of global temperature trends from satellite. *Science* **247,** 1558.

———, 1992: Precision and radiosonde validation of satellite gridpoint temperature anomalies, Part II: A tropospheric retrieval and trends during 1979–90. *J. Climate,* in press.

Stenger, P. J., and P. J. Michaels, 1992: Climatic change in mixed-layer trajectories over large regions. *Theor. Appl. Clim.* **44,** in press.

Stooksbury, D. E., 1991: Is climate change already giving us greater maize yeilds? *New Scientist* **132,** 48.

Wallace, H. A., 1920: Mathematical inquiry into the effect of weather on corn. *Mon. Wea. Rev.* **48,** 439–46.

Walsh, J. E., 1991: The Arctic as bellweather. *Nature,* **352,** 19–20.

Wang. W.-C., M. P. Dudeck, X.-Z. Liang, and J. T. Kiehl, 1991: Inadequacy of effective CO_2 as a proxy in simulating the greenhouse effect of other radiatively active gases. *Nature,* **350,** 573–77.

Warren, S. G., C. J. Hahn, J. London, R. M. Chervin, and R. L. Jenne, 1988: *Global Distribution of Total Cloud Cover and Cloud Type Amounts over the Ocean,* U. S. Department of Energy Publication DOE/ER-0406, 42 pp + Maps.

Washington, W. M., and G. A. Meehl, 1983: General circulation model experiments on the climatic effects due to a doubling and quadrupling of carbon dioxide concentration. *J. Geophys. Res.* **88,** 6600–6610.

REFERENCES

Washington, W. M., and G. A. Meehl, 1989: Climate sensitivity due to increased CO_2: Experiments with a coupled atmosphere and ocean general circulation model. *Clim. Dyn.* **2,** 1–38.

Weber, G.-R., 1990a: Tropospheric temperature anomalies in the northern hemisphere 1977–86. *Int. J. Climatol.* **10,** 3–19.

Weber, G.-R., 1990b: Spatial and temporal variation of sunshine in the Federal Republic of Germany. *Theor. Appl. Climatol.* **41,** 1–9.

Wiesnet, D., and M. Matson, 1980: U.S. Department of Commerce, *Environmental Data and Information Service Reports* **11** (1).

Wigley, T. M. L., 1987: Relative contributions of different trace gases to the greenhous effect. *Climate Monitor* **16,** 14–28.

Wigley, T. M. L., 1991: Could reducing fossil-fuel emissions end global warming? *Nature* **349,** 503–5.

Wood, F. B., 1988: Global alpine glacier trends, 1960s to 1980s. *Arctic and Alpine Research* **20,** 404–13.

Woodward, F. I., 1987: Stomatal numbers are sensitive to increases in CO_2 from pre-industrial levels. *Nature* **327,** 617–19.

About the Author

Patrick J. Michaels is associate professor of environmental sciences at the University of Virginia and senior fellow in environmental studies at the Cato Institute. He has served for 12 years as the Virginia State Climatologist and in 1987–88 was president of the American Association of State Climatologists. He has published numerous articles in journals of climatology, forestry, and meteorology. Michaels is a member of the American Meteorological Society, the American Association for the Advancement of Science, and Sigma Xi.

Cato Institute

Founded in 1977, the Cato Institute is a public policy research foundation dedicated to broadening the parameters of policy debate to allow consideration of more options that are consistent with the traditional American principles of limited government, individual liberty, and peace. To that end, the Institute strives to achieve greater involvement of the intelligent, concerned lay public in questions of policy and the proper role of government.

The Institute is named for *Cato's Letters*, libertarian pamphlets that were widely read in the American Colonies in the early 18th century and played a major role in laying the philosophical foundation for the American Revolution.

Despite the achievement of the nation's Founders, today virtually no aspect of life is free from government encroachment. A pervasive intolerance for individual rights is shown by government's arbitrary intrusions into private economic transactions and its disregard for civil liberties.

To counter that trend, the Cato Institute undertakes an extensive publications program that addresses the complete spectrum of policy issues. Books, monographs, and shorter studies are commissioned to examine the federal budget, Social Security, regulation, military spending, international trade, and myriad other issues. Major policy conferences are held throughout the year, from which papers are published thrice yearly in the *Cato Journal*. The Institute also publishes the quarterly magazine *Regulation* and produces a monthly audiotape series, "Perspectives on Policy."

In order to maintain its independence, the Cato Institute accepts no government funding. Contributions are received from foundations, corporations, and individuals, and other revenue is generated from the sale of publications. The Institute is a nonprofit, tax-exempt, educational foundation under Section 501(c)3 of the Internal Revenue Code.

CATO INSTITUTE
224 Second St., S.E.
Washington, D.C. 20003